NUREG-1764

Guidance for the Review of Changes to Human Actions

Final Report

Manuscript Completed: January 2004
Date Published: February 2004

Prepared by
J.C. Higgins*, J.M. O'Hara*
P.M. Lewis**, J.J. Persensky**, J.P. Bongarra**
S.E. Cooper**, G.W. Parry**

*Brookhaven National Laboratory
Energy Sciences and Technology Department
Upton, NY 11973-5000

Division of Inspection Program Management
Office of Nuclear Reactor Regulation
**U.S. Nuclear Regulatory Commission
Washington, DC 20555-0001
NRC Job Code Number Y6022

ABSTRACT

The U.S. Nuclear Regulatory Commission (NRC) is addressing the human performance aspects of changes to operator actions that are credited for safety, especially those involving changes in the licensing basis of the plant, e.g., using a manual action in place of an automatic action for safety system operations. This document provides guidance for reviewing changes to human actions that are credited for safety. In this document, the term human action and operator action are used synonymously because most of the types of actions discussed are performed by operations staff. The evaluation method uses a two-phase approach. The first phase is a screening analysis of the licensee's proposed modification and the affected human actions to assess their risk importance. A graded, risk-informed approach is used to determine the appropriate level of human factors engineering review. This approach can be accomplished for submittals by licensees that are either risk-informed or non-risk-informed. For a risk-informed submittal, the first phase has four steps: Use of Regulatory Guide 1.174 to determine the risk importance of the entire plant change or modification which involves the human action; a quantification of the risk importance of the human action itself; a qualitative evaluation of the human action; and, an integrated assessment to determine the appropriate level of human factors engineering review. The proposed human actions are assigned to one of three risk levels (high, medium, low) as a result of phase 1. The level of the human factors review in the second phase corresponds to these risk levels. In the second phase, human actions are reviewed using standard criteria in human factors engineering to verify that the proposed action can be reliably performed when required. Human actions in the high risk level receive a detailed human factors engineering review and those in the medium risk level undergo a less detailed one, commensurate with their risk. For human actions placed in the low risk level, there is a minimal human factors review or none. The NRC review of licensee submittals that involve changes to human actions is an iterative process. The final results of the human factors review will provide input to the Integrated Decision-making and to a Safety Evaluation Report.

CONTENTS

List of Figures

List of Tables

EXECUTIVE SUMMARY

The U. S. Nuclear Regulatory Commission (NRC) reviews changes in human actions (HAs) that are credited in plant safety analyses. Changes in credited action may result from a variety of plant activities such as: plant modifications, procedure changes, equipment failures, justifications for continued operations (JCOs), and identified discrepancies in equipment performance or safety analyses. Relevant considerations for review are described in NRC information notices and generic issues. Information Notice (IN) 91-18 (NRC, 1991) discusses the conditions under which manual actions may be used in place of automatic actions for safety-system operations. IN 97-78 (NRC, 1997) alerts licensees to the importance of considering the effects on human performance of such changes made to plant safety systems.

This document contains guidance to determine the appropriate level of human factors engineering review of human actions based upon their risk importance. A graded, risk-informed approach is used herein that is consistent with Regulatory Guide (RG) 1.174 Rev. 1 (NRC, 2002). Risk insights are used to determine the level of regulatory review the staff should perform. This approach can be accomplished for submittals by licensees that are either risk-informed or non-risk-informed. Human actions that are considered more risk-significant receive a detailed review, while those of less significance receive a less detailed one. In this document the term HA and operator action are used synonymously because most of the types of actions addressed are performed by operations staff. In keeping with RG 1.174, this guidance does not preclude other approaches for requesting changes to a plant's licensing basis or for requesting changes in HAs. Rather, this method of reviewing HAs is intended to improve consistency in regulatory reviews and decisions.

A two-phase evaluation is used. The first phase is a risk screening and analysis of the affected HAs identified by the licensee and a determination of their risk importance. The second phase is a human factors engineering review of those HAs found to be risk-important. Each is described below.

Risk Screening Process

A plant change may include changes to equipment, as well as to human actions. This approach focuses on the HA, while risk screening of equipment changes can be accomplished directly using RG 1.174. The risk screening for submittals by licensees that are risk-informed is described in Section 2.3 and for those that are non-risk-informed in Section 2.4.

A four-step screening process is used to locate the plant modification and its associated HAs in risk space, using guidance similar to that of RG 1.174. The first two steps are quantitative while the third is qualitative. The fourth step is an integrated assessment that considers the results of the first three steps and determines the appropriate level of human factors review. Essentially, plant modifications and their associated HAs can be categorized into regions of high, medium, and low risk. This categorization sets the level of graded human factors engineering review needed. Following are the important steps of this process.

Prior to submitting a change request to the NRC the licensee reviews a proposed plant change to identify HAs that are new actions, modified actions, or involve modified task demands. The licensee conducts a 10 CFR (Code of Federal Regulation) 50.59 evaluation for any changes that affect the licensee's Final Safety Analysis Report (FSAR). This evaluation may identify activities associated with the change, that require the NRC's review and approval before they are implemented.

For the risk-informed review, the licensee would make an initial screening risk calculation and submit this to NRC with a request for approval of the change. The first step of the risk screening evaluates the full modification including both equipment and HAs and conducted using the NRC's RG 1.174 directly. Risk calculations include the change in risk or core damage frequency (CDF) due to the full modification (ΔCDF_{mod}) that includes the HA. This assessment may determine that the full modification, including the HA, is in Region III (the low risk region). If so, then in some cases the only NRC review that is necessary, is to evaluate whether there is a valid technical basis for the low risk.

The risk screening calculations also consider whether the proposed change is permanent or temporary. If temporary, the screening includes considering the length of time that it will be in place. Then, the method assesses the integrated risk due to the change or modification over the time that it will be in place (or the integrated conditional core damage probability - ICCDP). Similar calculations would be performed for large early release frequency (LERF) where appropriate.

The second step of the risk screening focuses on the risk importance of the HA. This step uses both the Risk Achievement Worth (RAW) and the Fussell-Vesely (FV) risk importance measures to determine the risk significance of the HA. This step identifies the effect on risk of failing to perform the HA (using the RAW) and the relative contribution to risk of the action (using the FV). Uncertainty about human actions is treated by setting the HA failure probability to 1.0 for the action under review. This second step of the risk screening tentatively places the HA in one of three risk levels (high, medium, or low) for determining the level of human factors engineering review to be performed by NRC. The importance of the HA with respect to both CDF and LERF is then assessed. For HAs determined to be risk significant (i.e., in the high or medium risk levels), the intent of the detailed human factors engineering review is to verify that they can be successfully performed, when required, to limit the risk associated with failure of the HAs.

The third step of the risk screening is qualitative, and allows NRC to adjust the quantitative evaluation from Step 2, if necessary, considering factors that cannot be taken into account quantitatively in the PRA model. It includes several factors, such as personnel functions and tasks, design support for task performance, and performance shaping factors.

The fourth step is an integrated assessment that considers the results of the first three steps and determines the appropriate level of human factors review. If the action is verified to be in the low risk level, then the licensee's change likely would be permitted with either minimal or no further NRC human factors review. If the action is in the medium risk level then the NRC undertakes a moderate, top level human factors review. If the action is in the high risk level, then a more detailed review required, including human factors engineering, deterministic reviews, and risk analysis.

Human Factors Engineering Review

In this phase, the proposed HAs are reviewed to verify that they can be reliably performed when needed. Again, the details of the review are commensurate with the risk. There are three levels of risk and NRC review are presented. The review criteria are based on an adaptation of existing NRC review guidance for human factors, as found in NUREG-0800 (NRC, 2004a), NUREG-0711 (NRC, 2004), NUREG-0700, (NRC, 2002b), and IN 97-78 (NRC, 1997).

A Level I review is used for HAs in the high risk category. Changes in Level I require the most stringent review and include most of the review elements from NUREG-0711, the NRC's Human Factors Engineering Program Review Model.

The NRC conducts a Level II review for HAs in the medium risk category. While the guidance addresses similar topical areas as the Level I review, the extent of the staff review is notably less. The evaluation processes for this level are less prescriptive and provide greater latitude to both the licensee and the NRC reviewers for collecting and analyzing information. The Level II evaluation process addresses the following elements: General Deterministic Review Criteria; Analysis; Design of Human-System Interface (HSI), Procedures, Training; and Human Action Verification.

HAs in the low risk category receive only a limited Level III review by the NRC, generally limited to verifying that the action is in fact in Level III. Typically, no human factors engineering review is necessary. However, NRC reviewing may use a few review areas based on the results of Step 3 of the risk screening process.

Licensees may choose to use the Level II guidance to address human factors considerations for HAs that fall into Level III.

Final Decision on Acceptance of Human Actions

The NRC review of licensee submittals involving changes to HAs is an iterative process. That is, the final results of the human factors review will provide input to the Integrated Decisionmaking (see RG 1.174, Section 2.2.6), and may be documented in a Safety Evaluation Report.

The results of the different analyses are considered in an integrated manner (i.e., the decision is not driven solely by the numerical results of the risk assessment). This approach complements the NRC's deterministic approach, supports its traditional defense-in-depth philosophy, and takes into account traditional engineering, human factors engineering, and risk information. Both qualitative and quantitative analyses and information are used. The main factors considered in the decision process include the following:

- Change in CDF - The increase in CDF due to the modification (ΔCDF_{mod}).

- Change in LERF - The increase in LERF due to the modification ($\Delta LERF_{mod}$).

- Risk Importance Measures for the Human Action - The values of the Risk Achievement Worth (RAW) and Fussell-Vesely (FV) risk importance measures.

- Time and Integrated Risk - Risks integrated over the length of time that a temporary change will be in place.

- Human Factors - The basis that operators can perform the actions required for the modification, as determined by the human factors engineering review criteria used by NRC for the review.

- Deterministic Criteria - Satisfaction of the deterministic review guidance provided in Section 3.1 of the Level I review guidance or Section 4.1 of the Level II review guidance.

Additional or supplemental factors that could influence the acceptability of a change are also provided in the guidance document.

FOREWORD

This document provides NRC staff with guidance for reviewing changes to human actions in nuclear power plants. This document was developed (1) to consolidate into one document and to standardize review guidance that previously existed in several different documents, and (2) to make the review guidance more risk-informed. This guidance was specifically developed to review submittals for changes to credited human actions, such as plant changes that change the way credited actions are performed, such as reducing the time available and using manual action in place of automated action.

The review guidance has two phases. Phase 1 is a screening process that determines the level of human factors engineering (HFE) review in Phase 2. Three levels of HFE review (i.e., degree of detail and thoroughness) are possible: high, medium, and low.

The review guidance is appropriate for both risk-informed and non-risk-informed submittals. For a risk-informed submittal, Phase 1 has four steps: In Step 1, Regulatory Guide 1.174 is used to determine the risk importance of the entire plant modification that involves the human action. Step 2 is a quantification of the risk importance of the human action itself. Step 3 is a qualitative assessment of the safety significance of the human action. Step 4 is an integrated assessment of the safety significance of the human action.

For a non-risk-informed submittal, in Phase 1 the safety significance of the human action is determined without benefit of the risk inputs that are provided in a risk-informed submittal. In this case, there is no PRA information submitted by the licensee. The NRC staff performs a scoping type risk evaluation to estimate the risk importance of the change to the human action.

Phase 2 is the HFE review, which, as noted above, is divided into the three levels: high, medium, and low. The review criteria for these levels were adapted from NUREG-0711.

Following Phase 2, a decision as to the acceptability of the proposed change is made considering several decision factors, including: change in risk, human factors criteria, deterministic criteria, time a modification will be in place, and supplemental decision factors related to the unique characteristics of the situation.

This guidance is used by NRC to review changes in human actions, including changes that are credited in plant safety analyses. Changes in credited action may result from a variety of plant activities such as: plant modifications, procedure changes, equipment failures, justifications for continued operations (JCOs), and identified discrepancies in equipment performance or safety analyses. The results of the human factors review will provide input to Integrated Decisionmaking and to a Safety Evaluation Report.

Farouk Eltawila, Director
Division of Systems Analysis and Regulatory Effectiveness
Office of Nuclear Regulatory Research

PAPERWORK REDUCTION ACT STATEMENT

Any information collections referenced in this NUREG are covered by the requirements of
10 CFR Parts 50 and 51, which were approved by the Office of Management and Budget, approvals
numbers 3150-0011 and 3150-0151.

ACKNOWLEDGMENTS

Several individuals were involved in developing this document. James C. Higgins and John M. O'Hara, Brookhaven National Laboratory, were the principal contributors. Susan Cooper and Gareth W. Parry of the U. S. Nuclear Regulatory Commission updated and modified the risk screening methodology of Section 2. Paul M. Lewis and J. J. Persensky of the U. S. NRC, Office of Nuclear Regulatory Research, provided overall management in preparing this document. James P. Bongarra, Jr., U. S. NRC, Office of Nuclear Reactor Regulation, was the technical coordinator and a reviewer for this project. Staff of the Office of Nuclear Reactor Regulation who reviewed this document were Richard J. Eckenrode, Clare P. Goodman; Michael C. Cheok, Martin Stutzke and Michael Franovich. Nathan O. Siu and Erasmia Lois of the U. S. NRC, Office of Nuclear Regulatory Research also reviewed the document.

ACRONYMS

ADS	automatic depressurization system
AFW	auxiliary feedwater
ANS	American Nuclear Society
ANSI	American National Standards Institute
AOT	allowed outage time
ATWS	anticipated transient without scram
BNL	Brookhaven National Laboratory
BWR	boiling water reactor
CCDF	cumulative value of core damage frequency
CDF	core damage frequency
CFR	Code of Federal Regulations
CR	control room
DBE	design basis event
DEP	depressurization
DHR	decay heat removal
ECCS	emergency core cooling system
EOP	emergency operating procedures
FSAR	final safety analysis report
FV	Fussell-Vesely
FW	feedwater
GDC	general design criteria
HA	human actions
HEP	human error probability
HF	human factors
HFE	human factors engineering
HPCI	high pressure coolant injection
HRA	human reliability analysis
HSI	human-system interface
I&C	instrumentation and control
IC	isolation condenser
ICCDP	incremental conditional core damage probability
ICLERP	incremental conditional large early release probability
IMC	Inspection Manual Chapter
IN	information notice
IPE	individual plant examination
ISLOCA	interfacing systems LOCA
JCO	justification for continued operations
LB	licensing basis
LERF	large early release frequency
LOCA	loss-of-coolant accident
LOOP	loss of offsite power
MSLB	main steam line break
NEI	Nuclear Energy Institute
NPP	nuclear power plant
NRC	Nuclear Regulatory Commission
OER	operating experience review

PORV	power-operated relief valve
PRA	probabilistic risk assessment
PSA	probabilistic safety assessment
PSF	performing shaping factor
PWR	pressurized water reactor
RAI	request for additional information
RAW	risk achievement worth
RCIC	reactor core isolation cooling
RCP	reactor coolant pump
RCS	reactor coolant system
RG	regulatory guide
RIS	regulatory issue summary
RI	risk-informed
SBO	station blackout
SG	steam generator
SGTR	steam generator tube rupture
SLC	standby liquid control
SRA	senior reactor analyst
SROA	safety-related operator action
SRP	standard review plan
SRV	safety relief valves
SSC	structures, systems, and components
TA	task analysis
USQ	unreviewed safety question
V&V	verification and validation

1   INTRODUCTION

In Information Notice (IN) 91-18 (NRC, 1991), the U.S. Nuclear Regulatory Commission (NRC) discussed the conditions under which manual actions may be used in place of automatic actions for safety system operations. IN 97-78 (NRC, 1997) alerted licensees to the importance of considering the effects on human performance of such changes made to plant safety systems:

> The original design of nuclear power plant safety systems and their ability to respond to design-basis accidents are described in licensees' FSARs (final safety analysis report) and were reviewed and approved by the NRC. Most safety systems are designed to rely on automatic system actuation to ensure that the safety systems are capable of carrying out their intended functions. In a few cases, limited operator actions, when appropriately justified, were approved. Proposed changes that substitute manual action for automatic system actuation or that modify existing operator actions, including operator response times, that were not reviewed and approved during the original licensing review of the plant may raise the issue of an unreviewed safety question (USQ). Such changes must be evaluated under the criteria of 10 CFR 50.59 to determine whether a USQ is involved and whether NRC's review and approval are required before implementation... In the NRC staff's experience, many of the changes involving operator actions proposed by licensees do involve a USQ. (p. 3)

While it is recognized that 10 CFR 50.59 was updated to remove the USQ wording, the intent of IN 97-78 is still pertinent: that is, many changes to operator actions still need to be submitted to the NRC for review and approval in accordance with the revised 10 CFR 50.59.

The term "safety-related operator action" (SROA) is defined in (American National Standard Institute/American Nuclear Society) ANSI/ANS-58.8-1994:

> A manual action required by plant emergency procedures that is necessary to cause a safety-related system to perform its safety-related function during the course of any design basis accident (DBE). The successful performance of a safety-related operator action might require that discrete manipulations be performed in a specific order. (p.4)

The guidance presented in this document can be used to address SROAs, as well as other required operator actions.

Licensee requests for changes may involve human actions (HAs) as a result of such plant activities as:

- plant modifications
- procedure changes
- equipment failures
- justifications for continued operations (JCOs)
- identified discrepancies in equipment performance or safety analyses.

The licensee's request should include an evaluation of such changes in plant activities to determine their effect on HAs. The following types of HA effects may result from these changes:

- New actions - An action that was not previously performed by personnel, such as when an action formerly performed by automation is allocated to the operators.

- Modified actions - A change in the way actions were previously performed, such as through introducing new task steps (e.g., due to new system components, a modification to a component, or failed components), or new control and display devices for performing the action.

- Modified task demands - Rather than affecting the task steps themselves, a change in the plant may affect the task demands, such as the amount of time available or the overall environment for the task .

This document proposes using a graded, risk-informed approach in conformance with Regulatory Guide (RG) 1.174 (NRC, 1998) and provides guidance for reviewing the human performance aspects of changes to plant systems and operations. Risk insights are used to determine the level of regulatory review the staff should undertake. HAs that are considered more risk significant receive a detailed review, while those of less risk significance receive a less detailed one commensurate with their risk. In this document, the term HA and operator action are used synonymously because most of the types of actions discussed are performed by operations staff.

The evaluation method uses a two-phase approach. The first phase is a three-step screening process to locate the plant modification and its associated HAs in risk space, using guidance similar to that of RG 1.174. The first two steps of the risk screening process are quantitative, while the third step is qualitative. Essentially, plant modifications and their associated HAs are categorized into regions of high, medium, and low risk that determines the level of graded human factors engineering (HFE) review needed.

In the second phase, the HFE review, the HAs are examined. The intent of this phase is to verify that the proposed HA can be reliably performed when needed. The details of the review are commensurate with the risk. There are three levels of NRC review. A Level I review is used for HAs in the high risk category (see Section 3 of this document). It examines the licensee's planning, analysis, design activities, and verification and validation as related to the change. The review criteria are based on an adaptation of existing NRC review guidance for HFE found in NUREG-0800 (NRC, 2004a), NUREG-0711, (NRC, 2004), NUREG-0700, (NRC, 2002b), and IN 97-78 (NRC, 1997). This was accomplished by analyzing past cases reviewed by NRC (Higgins, et al., 1999). While HAs in the high risk area of Region I generally are not desired, there are certainly examples of such actions in plants today, such as the pressurized water reactor (PWR) emergency core cooling system (ECCS) switchover. Also, there may be extenuating circumstances in which the licensee can adequately justify a modification to add a Region I HA, e.g., if the change is temporary or if there are other changes that reduce the core damage frequency (CDF). Another important consideration is how well the licensee has addressed the HFE aspects of the modification.

HAs in the medium risk category receive a Level II review by the NRC. While the guidance covers the same topical areas as the Level I review, the extent of the review is notably less (see Section 4 of this document).

Finally, the third level is called low risk to indicate that the modification involves less risk than those in the high or medium levels. For HAs in the low risk category (Level III), the staff review generally would be limited to verifying of the technical basis that the action is, in fact, in Level III. Such a verification can be made by reviewing the licensee's analysis methods and risk results that show the placement of the action in that risk level. Typically, no human factors review is necessary. However, the NRC may address a few review areas based on the results of Step 3 of the risk screening process.

2

In keeping with RG 1.174, this guidance does not preclude the licensee from using other approaches for justifying changes to a plant's licensing basis or for requesting changes in HAs. Rather, this review method is intended to improve consistency in regulatory decisions in areas where the results of risk analyses are used to help justify regulatory action. RG 1.174 notes the that risk-informed principles, process, and approach provide useful guidance for applying risk information to a broader set of activities than plant-specific changes to a plant's licensing basis. This document was developed within the spirit of such applications.

The RG notes that the use of probabilistic risk assessment (PRA) technology should be increased in all regulatory matters to the extent supported by the state-of-the-art in PRA methods and data. Its application should complement the NRC's deterministic approach and support the NRC's traditional defense-in-depth philosophy. The NRC's review of HAs also considers this concept.

RG 1.174 notes that decisions on proposed changes are expected to be reached in an integrated fashion, considering traditional engineering and risk information. They may be based on qualitative factors as well as quantitative analyses and information. Thus, the approach presented herein also considers such qualitative factors, both in Step 3 of the risk screening, and in the final decision on acceptance of human actions.

The Commission has also noted that the regulatory process should become "risk-informed" as opposed to "risk-based" (Thadani, 1998, p.1). Thus, the approaches described retain some deterministic aspects, for example dealing with defense-in-depth, meeting existing regulatory requirements, and addressing the HFE aspects of the HAs.

The NRC review of risk-informed changes that involve changes to HAs is an iterative process. The final results of the human factors review provide input to the Integrated Decisionmaking (see RG 1.174, Section 2.2.6), and may be documented in a Safety Evaluation Report.

This guidance contributes to satisfying the NRC goals of (1) maintaining safety, (2) increasing public confidence, (3) increasing regulatory efficiency and effectiveness, and, (4) reducing unnecessary regulatory burden. By implementing the guidance in this document, the NRC will improve the regulatory process in three areas: through enhancing safety decision-making by using PRA insights; through more efficient use of agency resources; and through reducing unnecessary burdens on licensees. The use of risk insights by licensees in submittals that request plant changes to HAs will assist the staff in the disposition of such licensee proposals.

2   RISK SCREENING PROCESS

2.1   Licensee Change Requests Involving Human Actions

As discussed in Section 1 of this document, licensee requests for changes may involve human actions (HAs) resulting from a variety of plant activities. These changes may result in new modified human actions, or modified task demands. This section provides guidance on deciding the level of human factors engineering review required for such HA changes using a risk-informed screening process. This approach can be accomplished for submittals by licensees that are either risk-informed or non-risk-informed. The risk screening for submittals by licensees that are risk-informed is described in Section 2.3 and for those that are non-risk-informed in Section 2.4.

2.2   Overview of Screening Process

The screening process for determining the appropriate level for Human Factors (HF) review of HAs has two major pathways, based upon the type of information contained in the licensee request for change:

    1.   Risk-informed submittals, (Section 2.3) and
    2.   Non-risk-informed submittals (Section 2.4).

Regulatory Guide (RG) 1.174 and Chapter 19 of the Standard Review Plan (NUREG-0800) provide guidance for reviewing risk-informed (RI) licensee change requests. RG 1.174 and the Standard Review Plan (SRP) also provide guidance on when a risk-informed submittal is required from the licensee. For RI submittals, adequate risk information should be available to assist in deciding on the appropriate level of HF review. The methodology for making this determination is discussed in Section 2.3 below.

This document provides guidance in two areas for non-risk-informed (non-RI) licensee change requests. First, an assessment of the appropriateness of a non-RI submittal for the human action(s) is provided that is based upon the guidance, given in SRP Chapter 19 and RG 1.174, for when a RI-submittal is required. If the assessment determines that the licensee's submittal should be risk-informed, then the licensee should submit the appropriate risk information for the change to be considered. In this case, the submittal is now RI, so the guidance for RI submittals is used. Second, for those remaining submittals that are appropriately non-RI, risk inputs will not be available to assist in deciding upon the appropriate level of HF review. The basis for such requests might be a licensee safety analysis using 10 CFR 50.59. In this case, guidance for determining the appropriate level of HF review in the absence of explicit, plant-specific, risk inputs is based on Chapter 19 of the SRP, as well as more general human reliability and risk concepts provided herein. Section 2.4 provides guidance for non-RI licensee change requests.

Section 2.5 briefly describes the recommended levels of HF evaluating review that are the outputs of the RI and non-RI screening processes. The actual guidance for conducting the HF review is in Section 3 and 4.

2.3   Screening Process for Risk-informed Change Requests

The SRP, Chapter 19 and RG 1.174, are used to evaluate RI requests for changes to the plant's licensing basis (LB). This analysis is performed by a PRA analyst; however, a human factors review is one of the inputs to the integrated decision-making described in RG 1.174.

The guidance in this section supplements that in the SRP, Chapter 19, and RG 1.174 to support the determination of the appropriate level of HF review.

2.3.1   Overall Screening Approach for Risk-informed Requests

The overall screening approach for RI requests consists of the following four steps:

1. Assignment of the change request into Region I, II, or III using RG 1.174,
2. Calculation of importance measures for an HA involved in the change request,
3. Qualitative assessment of the safety significance of an HA involved in the request for licensing basis (LB) change, and
4. Integrated assessment of HA safety significance for determining the appropriate level of HF review (i.e., Level I, II, or II).

Sections 2.3.2 and 2.3.3 describe the first step in the screening process for RI requests. Sections 2.3.4 and 2.3.5 describe Steps 2 and 3 in the overall RI screening process for HA reviews.

The results of Steps 1, 2, and 3 are inputs to Step 4, the integrated assessment of HA safety significance. Section 2.3.6 describes how this integrated assessment is performed and the resulting evaluation of the appropriate level of HF review.

2.3.2   Assignment into RG 1.174 Acceptability Regions (Step 1)

An NRC PRA analyst reviews the licensee's submittal and assigns it into the RG 1.174 acceptability regions using the Regulatory Guide and SRP Chapter 19.

For this first step in the screening process, there are two possibilities:

*   The request is for a <u>permanent</u> LB change
*   The request is for a <u>temporary</u> LB change

If the request is for a permanent LB change, the guidance provided in RG 1.174 is used directly to determine into which of three "Regions" (i.e., Region I, Region II, and Region III) the overall request will be assigned. This assignment is one input to the integrated assessment of the acceptability of the proposed change that involves the HA.

RG 1.174 does not directly address requests for temporary LB changes. Section 2.3.3 following describes how Region assignments can be made for such requests for the purposes of HF review screening. This approach is similar to that previously used by NRC staff to evaluate requests for temporary changes.

Permanent changes may include equipment only, human actions only, or a combination of both. Equipment-only changes, which have no impact on HAs, are not within the purview of this approach. Changes that involve HAs only, HAs plus equipment changes, or equipment changes that affect HAs should be evaluated using RG 1.174 as the first step.

RG 1.174, Figure 3, Acceptance Guidelines for Core Damage Frequency (CDF) and Figure 4, Acceptance Guidelines for Large Early Release Frequency (LERF) have three risk regions (Regions I, II, & III), the most risk significant being Region I. These figures are reproduced here for convenience and illustration.

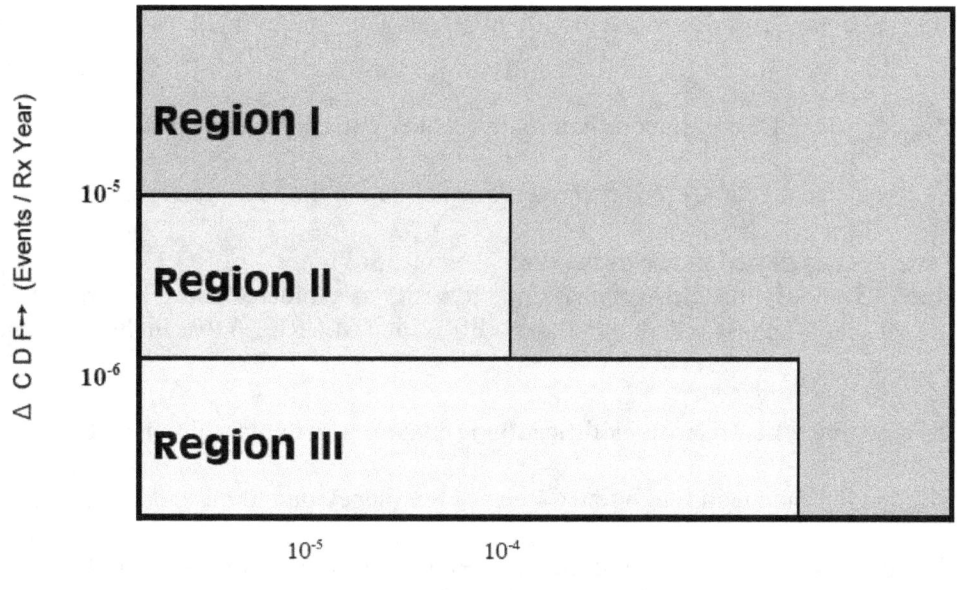

Figure 2.1 Acceptance guidelines for Core Damage Frequency (CDF)
(from RG 1.174 Figure 3)

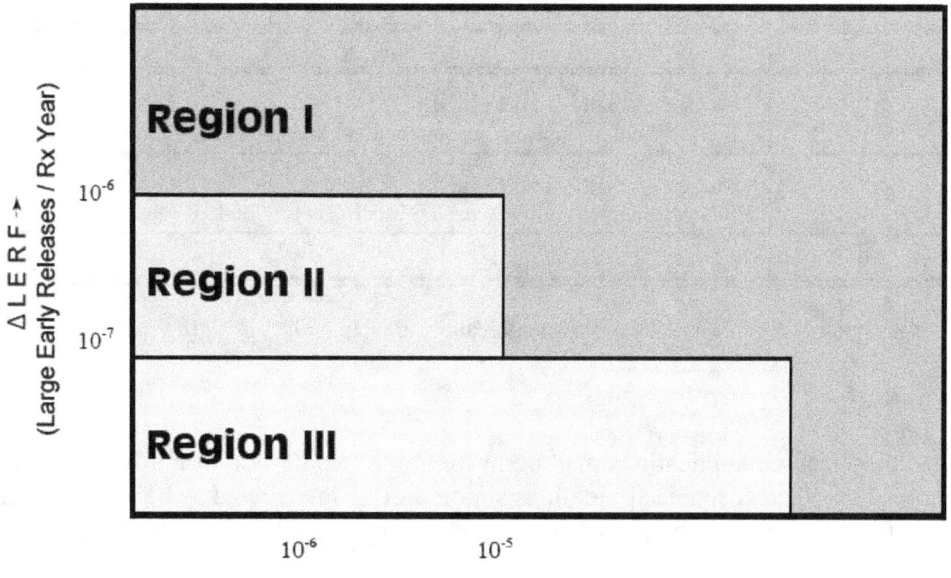

Figure 2.2 Acceptance guidelines for Large Early Release Frequency (LERF)
(from RG 1.174 Figure 4)

The change in core damage risk due to the modification (ΔCDF_{mod}) that includes the new human action is defined as

$$\Delta CDF_{mod} = [\text{new CDF (with modification in-place)} - \text{current baseline CDF}]$$

where, ΔCDF_{mod} is the change in Core Damage Frequency due to the modification.

This value of ΔCDF_{mod} is placed in one of the three Regions of Figure 3 of RG 1.174, for Step 1 of the screening method. Similarly, the change in risk due to LERF is evaluated using Figure 4 of RG 1.174. The change in large early release risk due to the modification ($\Delta LERF_{mod}$) that includes the new human action is defined as

$$\Delta LERF_{mod} = [\text{new LERF (with modification in-place)} - \text{current baseline LERF}]$$

where, $\Delta LERF_{mod}$ is the change in Large Early Release Frequency due to the modification.

This value of $\Delta LERF_{mod}$ is placed in one of the three Regions of Figure 4 of RG 1.174, for Step 1 of the screening method. The Region for Step 1 of the screening is the most conservative of Figure 3 or Figure 4 of the RG. If it is in Region I, the NRC will likely (but not definitely) disapprove the modification; this is discussed in more detail below. If the modification is in Region II or III, then the reviewer proceeds to Step 2 of the screening process, as summarized in Table 2.1 below.

Table 2.1 Action on completion of step 1

Step 1 Results	NRC Review Action
Region I (Equip. + HA)	- Change generally not permitted. - If change not disapproved at Step 1, go to Step 2 of screening.
Region I (HA only)	- Change generally not permitted. - If change not disapproved at Step 1, perform Region I HF review.
Region II or III	- Go to Step 2 of screening

Note: When using this Table for a temporary modification, use Region I^{intgr}, II^{intgr}, III^{intgr}.

Modification in Region I for Step 1

RG 1.174 notes that licensee applications that are in Region I "would not normally be considered." However, it notes that "the acceptance guidelines should not be interpreted as being overly prescriptive." There may be extenuating circumstances where they would be considered and approved, including these reasons:

- unquantified benefits that are not reflected in the quantitative risk results
- compensatory measures are proposed to counter the impact of major risk contributors.

If a Region I modification includes a combination of both equipment and HAs, NRC may reject the overall modification. Then, no further screening is necessary. If such a modification is not rejected, the screening proceeds to Step 2 to evaluate the risk significance of only the HA portion of the modification.

8

If a Region I modification has only HAs, the NRC may likewise reject it. If the change is not rejected, the reviewer proceeds directly to the Level I human factors engineering review. Since the HA itself is very risk significant, there is no need to perform Steps 2 & 3 of the screening; the NRC will review the HA using the guidance for Level I human factors engineering review in Section 3 of this document.

Modification in Region II for Step 1

If the overall modification is in Region II, it is still possible that the HA may undergo either a Level I or Level III human factors engineering review. Therefore, Steps 2 and 3 of the screening process must be performed.

Modification in Region III for Step 1

In the case where a Region III modification includes a combination of both equipment and HAs, there may be equipment improvements that cause a decrease in CDF and that may be masking the risk significant contribution of the HA. That is, though the overall modification is not risk significant, the HA may be when considered by itself. Therefore, the reviewers must undertake Steps 2 and 3 of the screening.

If a Region III modification includes only HAs, Steps 2 and 3 of the screening method still are performed. This is because the Step 1 risk calculation is based on the base-case value of human error probability (HEP) for the HA. Further, there may be a situation where a licensee is replacing a demonstrated reliable automatic component with a presumed reliable HA (with a low HEP). Step 2 will evaluate this and other possibilities.

2.3.3 Human Factors Screening Approach for Temporary Changes (Step 1)

Licensee requests for temporary changes often involve HAs that provide compensatory measures to offset the calculated increase in risk. For example, the request may propose substituting HAs for automatic equipment that is temporarily inoperable and cannot be restored within the time interval required by the plant technical specifications.

For temporary changes, the calculated risk increase (or decrease) must consider the time that the modification will be in place. RG 1.174 generally applies to all changes (both permanent and temporary) to the licensing basis of a plant. However, the acceptance guidelines in RG 1.174 that result in assignments to Region I, II, or III do not explicitly address the duration of a temporary change.

To support the screening process for temporary change requests, the guidance in RG 1.174 is supplemented to make assignments into Regions I, II, or III. The guidance following describes a method the PRA analyst uses to quantitatively evaluate, in an integrated fashion, both the increase in risk and the length of time of increased risk. For this application the Regions are named: Region I^{intgr}, II^{intgr}, III^{intgr} (shown in Fig. 2.3 and 2.4).

The risk calculated by a PRA can be expressed in several ways: as an instantaneous value (often calculated for configuration risk management purposes); an average value of CDF over a reactor year (the most common value cited); or, a cumulative value of core damage frequency (CCDF) computed over a defined time. The CCDF can be calculated accurately using statistical techniques. A simplified method of viewing the cumulative or integrated risk is to multiply the CDF by time. This gives reasonable results for the type of screening review the NRC is performing for HAs that are risk-important.

Thus, equations for integrated risk can be written as follows:

$$\text{Integrated } \Delta\text{CDF Risk (mod)} = \Delta\text{CDF}_{mod} \times \text{time (mod)} = \text{ICCDP, or}$$

$$\text{Integrated } \Delta\text{LERF Risk (mod)} = \Delta\text{LERF}_{mod} \times \text{time (mod)} = \text{ICLERP,}$$

where, Integrated Risk (mod) is the integrated risk due to the modification over the time that the change or modification is to be in place, expressed as CDF or LERF;
time (mod) is the length of time that the change or modification is to be in place; ICCDP is the incremental conditional core damage probability, and ICLERP is incremental conditional large early release probability.

The value of Integrated CDF Risk (mod) can be roughly interpreted as the change in the expected core damage probability in the plant over the time period due to the modification. RG 1.177 also uses this concept of integrated risk, where the Integrated CDF Risk is the incremental conditional core damage probability (ICCDP) and the Integrated LERF Risk (mod) is the incremental conditional large early release probability (ICLERP). Incremental in this situation refers to the incremental increase in risk over the time period for the modification.

To support the screening process used in determining the level of HF reviews, acceptance criteria similar to those in RG 1.174 were developed. They were adapted from Section 2.4 of RG 1.177 that addresses the acceptability of integrated risk over periods when equipment is out-of-service for the allowed outage time (AOT). In Section 2.4 of RG 1.177, an acceptability limit of 5E-7 per reactor-year for ICCDP is considered a small risk increase for a single Technical Specification AOT. This 5E-7 value is chosen as the boundary between Regions II^{intgr} and III^{intgr} for ICCDP. The selected boundary for Regions I^{intgr} and II^{intgr} is 5E-6 events per reactor-year, an increase of one order-of-magnitude. These boundary values result in Figure 2.3 that shows the acceptability limits for ICCDP in terms of Regions I^{intgr}, II^{intgr}, and III^{intgr}

Similarly, RG 1.177 uses 5E-8 events per reactor-year for the limit on a small LERF increase. This value was adopted for the boundary between Regions II^{intgr} and III^{intgr} for ICLERP, while 5E-7 is used for the Region $\text{I}^{intgr}/\text{II}^{intgr}$ boundary. Figure 2.4 shows the acceptability limits of ICLERP resulting from these choices.

Similar to the approach in RG 1.174, both Figures 2.3 and 2.4 are used by the PRA analyst to determine acceptability if the modification affects LERF. If LERF is not affected by the change, then Figure 2.3 alone will suffice.

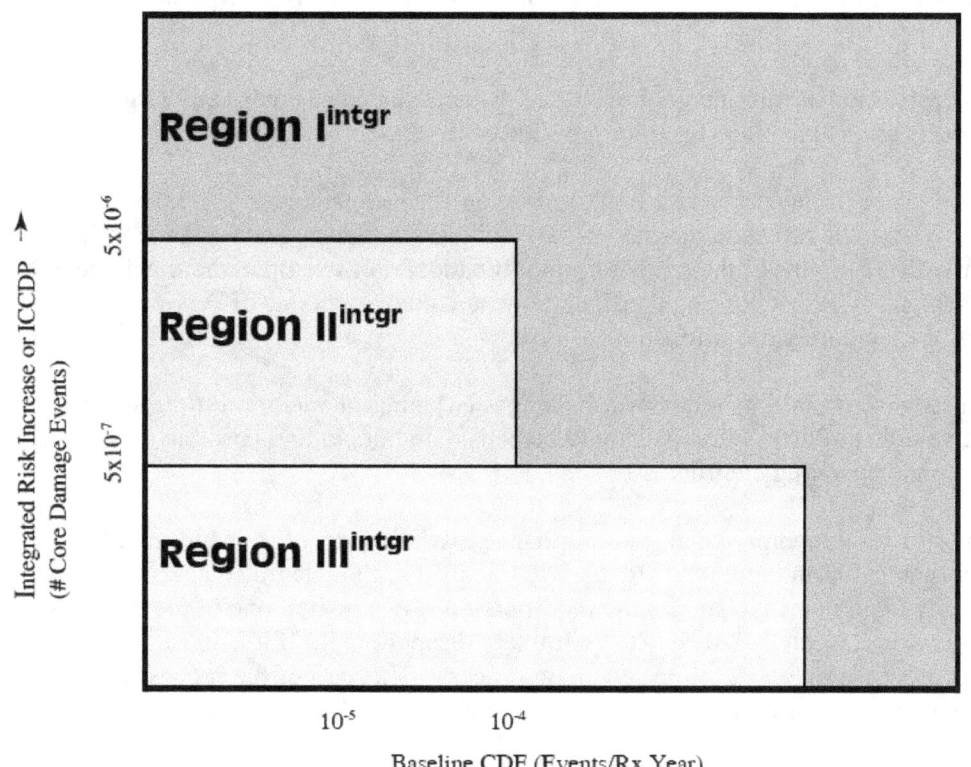

Figure 2.3 Guidelines for integrated risk increase -ICCDP
(Product of ΔCDF_{mod} and Time)

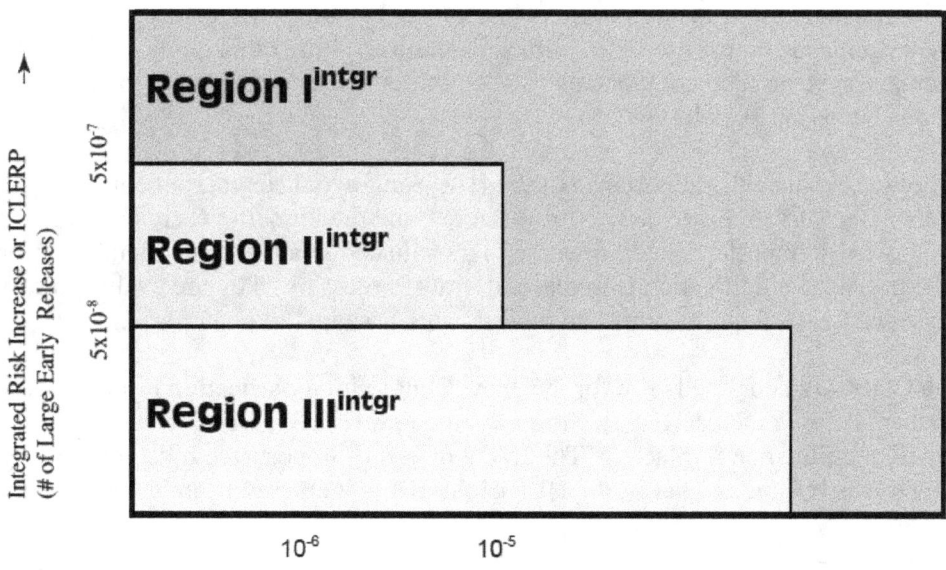

Figure 2.4 Guidelines for integrated risk increase - ICLERP
(Product of $\Delta LERF_{mod}$ and Time)

These are potential concerns about using integrated risk measures to address requests for temporary changes:

- Strict interpretation of this approach would allow acceptance of potentially large increases in risk if the modification is in place for a short period of time.

- For larger values of integrated risk (e.g., in Region I^{intgr}), consideration should be given to the potential synergistic effect on the risk of CDF spikes due to changes in plant configuration together with the effects of the temporary modification. For example, there may be a need to impose temporary restrictions on configurations and equipment out-of-service during the time period of the temporary modification.

- If a licensee makes multiple requests for changes (whether in one or multiple submittals), the cumulative effect of these requests should be considered, including possible dependencies and changes to the operating environment.

The result of this step for temporary changes is similar to that for permanent changes: a Region I^{intgr}, II^{intgr}, or III^{intgr} assignment is an input made by the PRA analyst to the integrated assessment discussed in Section 2.3.6. As with permanent changes, the modification is placed into one of the three risk regions. If the modification is in Region I^{intgr}, the NRC will likely disapprove it. If it is in Region II^{intgr} or III^{intgr}, then the reviewer proceeds to Step 2 of the screening process. This part of the process is similar to that for permanent changes, and is summarized in the preceding Table 2.1.

2.3.4 Calculation of Importance Measures for Human Actions (Step 2)

In Step 2 of the screening process for RI requests, importance measures are calculated to assess the importance of HAs involved in the LB change request. A PRA analyst performs this step, and the results are inputs in the integrated decision-making in Step 4. Chapter 19 of the SRP and RG 1.174, especially Appendix A, provide guidance on the use of importance measures. While this guidance is written primarily for structures, systems, and components (SSCs), the same principles apply to, and should be used for, calculations involving HAs.[1]

Step 2 addresses how to evaluate the importance of the HA, using two different, but complementary, risk importance measures: the Risk Achievement Worth (RAW), and the Fussell-Vesely (FV) importance measure. Both of these risk importance measures are first evaluated relative to the plant's new baseline CDF, assuming the proposed modification is in place. Next, if necessary, they are evaluated relative to the plant's new baseline LERF with the proposed modification assumed to be in place.

The "new baseline CDF" is a shortened term for the "new CDF (with modification in-place)" used in the previous section when defining the ΔCDF_{mod}. Similarly, the new baseline LERF is a shortened term for the "new LERF (with modification in-place)." The RAW measures importance by computing the increase in CDF when the HA fails. That is, the HEP of the HA is increased from its base case value to

---

[1] Chapter 19 of the SRP uses standard PRA terminology. Consequently, the SRP defines the basic event used to represent human actions as a "human failure event" representing failures of functions, systems, or equipment that are caused by human actions (or lack of action). As such, human failure events may represent more than one action. However, in this report, the term "human action or HA" will continue to be used, even when "human failure event" is the more appropriate term.

1.0 and the overall CDF is re-computed. Then, to compute the RAW, a ratio or a difference of the new higher CDF to the baseline CDF is taken. The more common ratio method of expressing RAW is used here. The RAW importance measure was defined and discussed in NUREG/CR-3385 (Vesely, et al., 1983) and (Lambert, 1975). One equation for the ratio value of the RAW for HA "x" is

$$RAW (x) = (CDF \text{ with x set to } 1.0) / CDF_{new\ BL}$$

A high RAW value means that failure of the HA results in a risk significant situation. Thus, the HA's reliability should be verified by a thorough human factors engineering review.

The FV importance measure represents a different way of expressing risk significance than RAW and is included to obtain a more robust evaluation of risk importance. FV is defined as the CDF of core damage cutsets (or sequences) that contain the HA in question, divided by the total CDF. This is expressed for HA "x" in the following equation:

$$FV(x) = \sum CDF \text{ of cut sets containing x} / CDF_{new\ BL}$$

If FV is high, the HA contributes to a relatively large portion of risk. Thus, for defense-in-depth purposes, the HA's reliability should be insured by a thorough human factors engineering review.

Figures 2.5, 2.6, 2.7, and 2.8, respectively, provide guidance on the level of human factors engineering review based on the calculated RAW and FV for the new baseline CDF and LERF. The curves delineating the boundaries between the different levels of review are related to the definitions for Regions I, II, and II given in RG 1.174, as discussed below (and in more detail: Higgins, et al. (2002), and in the response to public comments on NUREG-1764).

CDF Importance Measure Evaluation of HA

Figures 2.5 and 2.6 show the Level assignments for RAW and FV analogous to the RG 1.174 Regions. The term Level was chosen to represent an amount of risk due to the HA and the corresponding amount of human factors engineering review. The levels are distinct and different from the RG 1.174 Regions. We note that these levels are not related to the three PSA Levels of CDF, source term releases from containment, and offsite consequences. RAW and FV values, which should be computed for the HA being evaluated, together with the new baseline CDF, will determine where on the Figures the HA is placed and which level of human factors review is assigned by Step 2. If an HA falls very close to the dividing line between two levels, then the reviewer may want to use the qualitative criteria in Step 3 to finally decide on the level of review for the HA (in Steps 3 & 4 of the screening process).

The curve delineating the split between the Level I and Level II areas of Figure 2.5 is roughly based on a CDF of 1 E-4 core damage events per reactor-year, the subsidiary objective of the Commission's safety-goal policy statement. Performance deficiencies associated with actions in the Level I area would generally be colored as Red in the new Reactor Oversight program. Similarly, the curve delineating the split between the Level II and Level III areas of Figure 2.5 are roughly based on a CDF of 1 E-5 core damage events per reactor-year. Performance deficiencies associated with actions in the Level II area would generally be colored as Yellow, those in the Level III area would be colored as White or Green. Figure 2.6 and the FV importance measure were added to have a second, different risk importance measure that would add robustness to the method. This measure addresses HAs that may not have a high RAW value (e.g., due to a relatively high base case HEP), but which contribute notably to the base- case CDF.

13

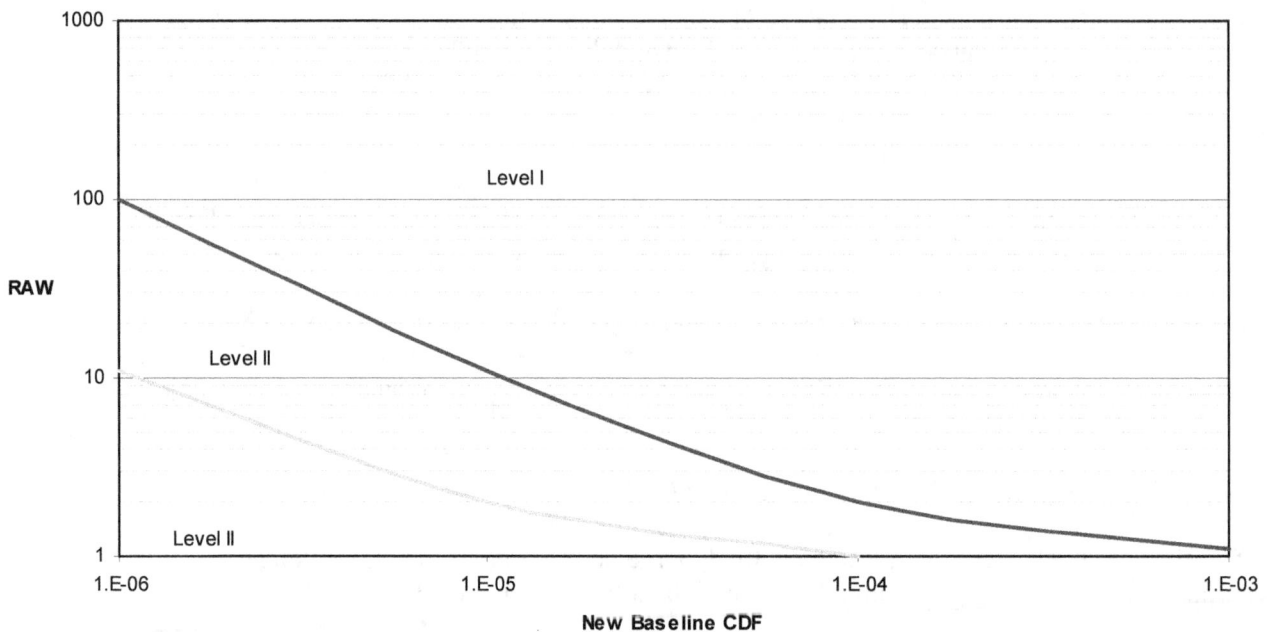

Figure 2.5 RAW vs. new baseline CDF

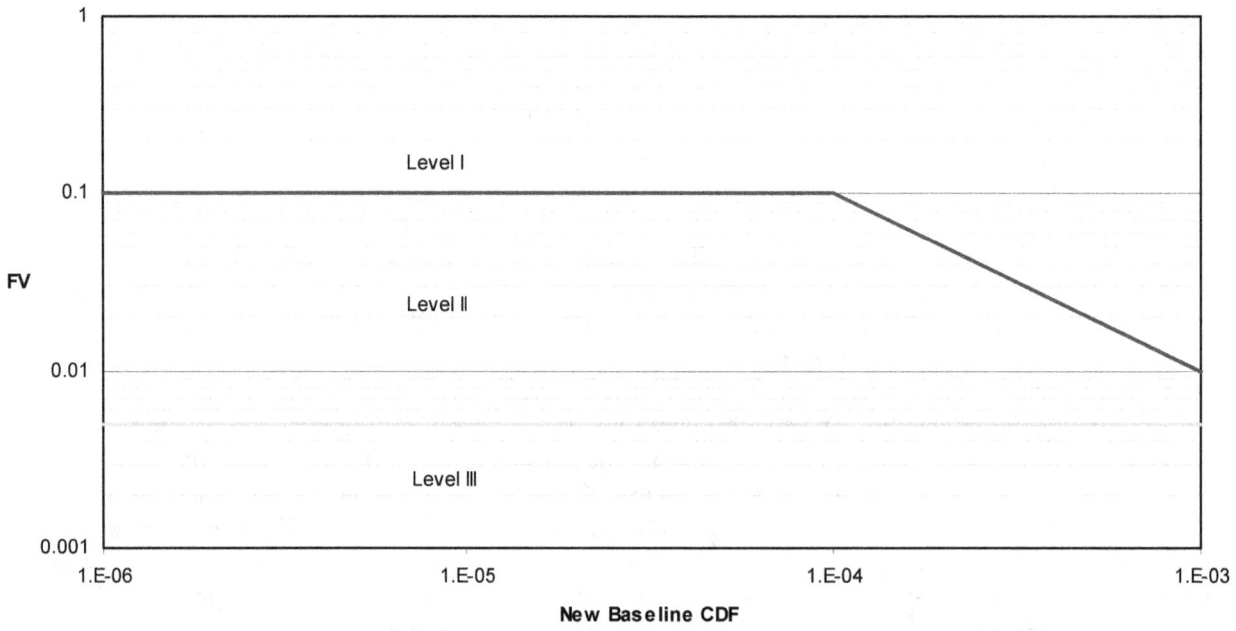

Figure 2.6 FV vs. baseline CDF

14

LERF Importance Measure Evaluation of HA

Next, the PRA analyst determines if the HA requires a separate LERF importance measure evaluation. In general, the default would be that it is not necessary to separately perform a LERF evaluation in this step because

- most HAs affect primarily CDF and the LERF evaluation would not yield a different risk-review level;

- LERF importance measures are not routinely calculated, while the CDF importance measures are; and,

- some PSA Level II models are not structured to support such calculations of LERF importance measures.

If the PRA analyst judges that an HA does affect the LERF calculations differently than the CDF calculations and that a LERF evaluation should be done, then the approach described immediately below can be used.

The method will use a LERF RAW importance measure, designated as RAW (L) and LERF Fussell-Vesely importance measure designated as FV (L). These measures are analogous to the RAW and FV for the CDF calculations used above and are being applied by the industry for other regulatory purposes. They are defined as follows:

$$\text{RAW (L) (x)} = (\text{LERF with x set to } 1.0) / \text{LERF}_{\text{new BL}}$$

$$\text{FV (L) (x)} = \sum \text{all LERF cutsets (or sequences) containing x} / \text{LERF}_{\text{new BL}}$$

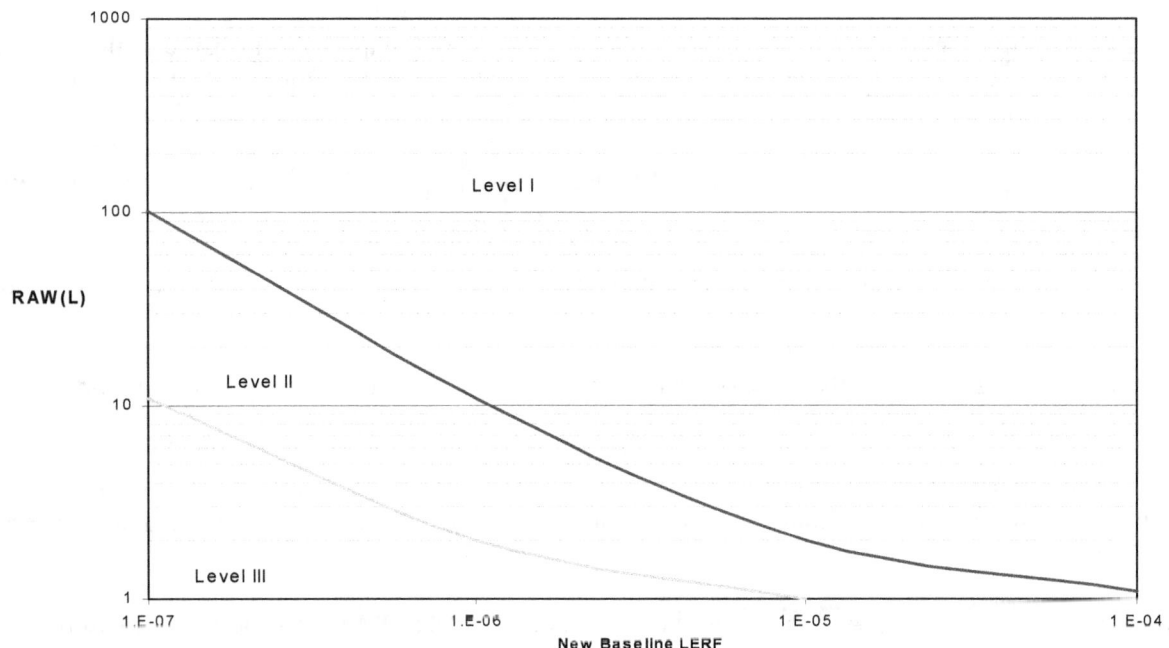

Figure 2.7 RAW (L) vs. new baseline LERF

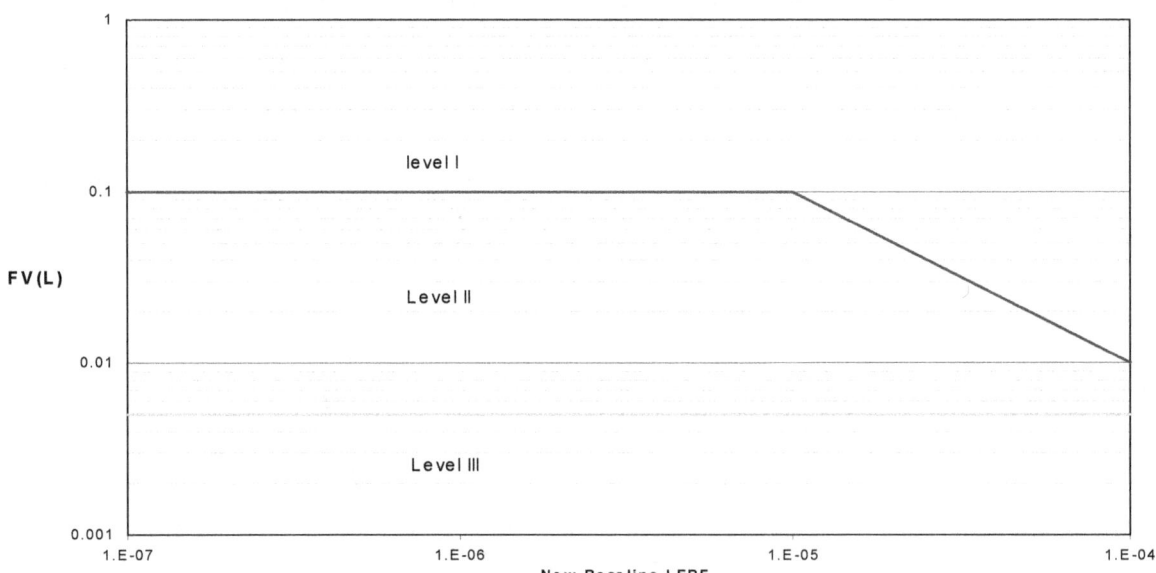

Figure 2.8 FV (L) vs. new baseline LERF

16

Figures 2.7 and 2.8 show the levels of review for RAW (L) and FV (L). They were adapted from Figures 2.5 and 2.6 by adjusting the values of the baseline LERF on the x-axis by one order-of-magnitude to account for the fact that LERF values and acceptance criteria generally are one-order-of magnitude less than CDF values and acceptance criteria.

The reviewer computes the RAW (L) and FV (L) values for the HA. This, together with the new baseline LERF, will determine the appropriate level of its HF review. If an HA falls very close to the dividing line between levels of review, then the reviewer may want to use the qualitative criteria in Step 3 below to finally decide on the level of review for the HA (in Steps 3 & 4 of this screening process).

After the regions for both RAW and FV are determined from Figures 2.5 and 2.6, (plus 2.7 and 2.8 if used) the HAs should be placed in the most conservative or highest risk region of the two figures (or four figures, if all were used). If the licensee undertook the calculation and placement on the Figures, the results should be submitted to the NRC for verification and use.

Multiple Human Actions

A particular plant modification may encompass multiple HAs some of which may be dependent. Also, the PRA modeling techniques for HAs vary, so that a given activity may be modeled as one HA or several. The screening evaluation should consider all of the HAs involved with the modification. Any dependent HAs should be aggregated together. That is, when computing the RAW and FV, include all HAs or all aspects of the one HA. This will give the full importance of the operator actions associated with the particular modification. Any HAs that are not dependent can be screened separately.

Consideration of Quality of PRA and Uncertainties

When evaluating the importance measures associated with the HA, careful attention should be paid to the quality of the PRA and the uncertainties that can affect the allocation. Its quality should be adequate to support the assessment of the risk significance of the HA. This may be demonstrated by applying an appropriate PRA Standard as endorsed by the NRC staff. A regulatory guide developed from DG-1122 soon will be issued for trial use. It will endorse the ASME PRA Standard and the industry's peer-review process (NEI-00-02). Also, the PRA analyst should make a judgment about the uncertainty of the HEP. If the HEP is too high or too low due to uncertainty or poor modeling, this will affect both the RAW and FV measures, but in opposite directions. Thus, Step 2 of the method tends to be robust in preventing uncertainty in the HEP from affecting the assignment of the level of review. As discussed in Appendix A of RG 1.174, uncertainties in parameter values can impact the assessment of the risk significance of a basic event when using importance measures. The assessment of the risk significance of an HA may be tested against the uncertainties in the assessment of the corresponding HEPs by performing appropriate sensitivity studies, varying the HEP through its range of uncertainty, as, for example, characterized by the 90% confidence interval. The final assessment should be conservative.

Further, if there are judged to be dependent HAs that were not properly modeled in the HRA and if the reviewer was unable to adequately address them (as discussed above), then increasing the level of human factors review of the set of dependent HAs should be considered.

There also may be cases when a lessening of defense-in-depth or safety margin is associated only with an HA. Then, an increase to the level of HF review would be appropriate. (Those cases associated with equipment should have been addressed already by reviews made with SRP, Chapter 19.)

2.3.5 Qualitative Assessment of Human Action Safety Significance (Step 3)

In Step 3 of the screening process for RI requests, a qualitative assessment is made of the HAs associated with the change request. This step will likely involve input from both PRA and HF analysts. The purpose of this qualitative assessment is to assess factors that cannot or may not have been addressed quantitatively in Steps 1 and 2. The results of this Step 3 may be recommendations for adjusting the level of HF review determined previously. These recommendations will be inputs to the integrated decision-making process (Step 4) for determining the appropriate level of HF review described in Section 2.3.6.

The results produced in the qualitative assessment step vary, depending on the specific factors involved and the analyst's assessment. In all cases, the results consist of recommendations for modifying the assignment of level of review based upon the risk calculations of Steps 1 and 2. These results can be either:

No change	No recommendations for changing the level of HF review from Steps 1 & 2 result from the qualitative assessments.
Elevate One Level	The results from the qualitative assessment indicate that the initially determined level of HF review in Steps 1 & 2 should be adjusted to a higher level.
Reduce One Level	The results from the qualitative assessment indicate that the initially determined level of HF review in Steps 1 & 2 should be adjusted to a lower level.

The qualitative factors to consider when making the assessment are identified in Section 2.3.5.1. Section 2.3.5.2 discusses the way those factors are used to adjust the level of HF review.

2.3.5.1 Factors Used in the Qualitative Assessments

Three types of qualitative assessment are used:

1. Personnel functions and tasks.
2. Design support for task performance.
3. Performance shaping factors.

The reviewer should determine whether the factors associated with each type of assessment are applicable to the particular HA.

Personnel functions and tasks

This type of qualitative assessment examines the potential effects of the request for change on operator tasks and the functions that they perform, under five major categories.

- *Operating Experience* – Does the requested change adversely affect the performance of an action that previously was identified as problematic based on experience/events at that plant or plants of similar design?

- *New actions* - Does the requested change introduce new human actions? Are the new human actions associated with new responsibilities for the success of safety functions (or additional actions associated with existing responsibilities)?

- *Change in Automation* - Has the requested change given personnel a new functional responsibility that they previously did not have and which differs from their normal responsibilities? For example, are operators now required to take an action in place of a previously automated one? Consider the example of simply being required to open a valve that previously was automatically operated, and where the action required to do so is similar to other valve-opening operations with which the operators are familiar. This would not be a sufficient change (in and of itself) to warrant a "yes" to this question when considering task complexity. However, there may be increased workload if the aggregate of added actions is judged to be excessive; this may warrant a "yes."

- *Change in Tasks* - Has the requested change significantly modified the way in which personnel perform their tasks (e.g., making them more complex, significantly reducing the time available to perform the action, increasing the operator workload, changing the operator role from primarily "verifier" to primarily "actor")? In this case, operators do not have a new functional responsibility; instead, the way that they perform their current functional responsibilities has significantly changed and is different from what they usually do.

- *Change in performance context* - Has the requested changed created, in some way, a new context[2] for task performance? Or, does the change identify a previously unrecognized context? Or, does the request address a context previously not modeled or considered? If so, what are the important differences in context (e.g., different plant mode, plant behavior, timing of plant symptoms)?

Design support for task performance

This type of qualitative assessment addresses how well the performance of the HAs is supported (e.g., with job aids).

- *Change in HSIs* (human-system interfaces) – Has the requested change significantly changed the HSIs used by personnel to perform the task? For example, are personnel now performing their tasks at a computer terminal where previously they were performed at a control board with analog displays and controls?

- *Change in Procedures* – Has the requested change significantly changed the procedures that personnel use to perform the task, or is the task not supported by procedures?

- *Change in Training* – Has the requested change significantly modified the training, or is the task not addressed in training?

---

[2] Here, context is defined as the overall performance environment, including plant conditions and behavior that, for example, affect the time available for the operator response and the effectiveness of job aids under these conditions that lead to the assessment of performance shaping factors.

Performance shaping factors

This type of qualitative assessment addresses four performance shaping factors.

- *Changes in Teamwork* - Has the requested change significantly changed the team aspects of performing an action. For example, (1) is one operator now performing the tasks accomplished by two or more operators in the past? (2) is it now more difficult to coordinate the actions of individual crew members? or, (3) is task performance more difficult to supervise after the modification?

- *Changes in Skill Level of Individuals Performing the Action* - Has the requested change kept the same HA but made it necessary for an individual who is less trained and has lower qualifications to take the action than was the case before the modification?

- *Change in Communication Demands* - Has the requested change significantly increased the level of communication needed to perform the task? For example, must an operator now communicate with other personnel to perform actions that previously could be taken at a local panel containing all necessary HSIs?

- *Change in Environmental Conditions* - Has the requested change significantly increased the environmental challenges (such as radiation, or noise) that could negatively affect task performance?

2.3.5.2   Adjusting the Level of Human Factors Review

Reduce the Level of HF Review

After considering the factors identified in Section 2.3.5.1, the analyst should consider reducing the level of HF review if the HA has the following characteristics.

- The answers are "no" to most of the questions posed by the analysis of the qualitative factors. This indicates that there is no change or very little change to the HA from the modification. One "yes" answer should not necessarily preclude a reduction in the level of the review, unless it is a "yes" to a significant question.

- The action is well defined and, based on the information presented in the submittal, the analyst is confident that it can be easily performed (e.g., it is clear when to perform the action, there are clear procedures, there is sufficient time and staff available, and the action is similar to those routinely taken.

The level of reviews should be reduced by only one level, e.g., Level II to Level III. Additionally, when the review is reduced to Level III, the following criteria taken from Chapter 19 of the SRP (Appendix C.2) should be used to verify that the SSCs or human (operator) actions are of low safety significance:

- The human (operator) action does not relate to the performance of a safety function or a support function to a safety function, or does not complement a safety function. The human (operator) action does not support (e.g., involve dependencies with) other operator actions that are credited in PRAs for either procedural or recovery actions.

- The failure of the human (operator) action will not result in the eventual occurrence of a PRA initiating event.

- The human (operator) action is not required in maintaining barriers to the release of fission products during severe accidents.

- The failure of the human (operator) action will not unintentionally release radioactive material, even in the absence of severe accident conditions.

If any of the above criteria are not satisfied, then re-elevation to a Level II HF review is recommended.

Elevate the Level of HF Review

The level of HF review *may* be increased if "yes" was the answer to any of the questions posed by the qualitative factors. The analyst should consider the potential effect of the "yes" factors. If such responses are obtained for many factors, the level of review of the HA should probably be increased. If a "yes" response is received for only one or two factors, then the analyst should consider whether the information leading to the "yes" response is sufficient to warrant elevating the level of review.

No Adjustment to the Level of HF Review

If the considerations of qualitative factors above did not decrease or increase the level of review, no change should be recommended on the basis of Step 3.

Result of Step 3

Thus, the final result of Step 3, Qualitative Assessment of HA Safety Significance, is one of three decisions that are used in Table 2.2 of Step 4: No Change, Elevate one Level, and Reduce one Level.

2.3.6   Integrated Assessment of Human Actions Safety Significance for Risk-informed Requests (Step 4)

This section provides guidance on how to integrate the results from Steps 1 through 3 of the screening process for risk-informed, licensing basis, change requests. This step is performed by both PRA and HF analysts. This guidance is intended to be conservative while still providing an opportunity to reduce the level of resources spent on HF review on the basis of risk and risk insights. This guidance is not intended to be prescriptive.

Table 2.2 summarizes the guidance for determining the appropriate level of HF review. This table shows the various possible results from Steps 1, 2, and 3, along with recommendations for the appropriate level of HF review. The results from Step 1 are the acceptability regions from RG 1.174. The results from Step 2 are the initial determination of the level of HF review from calculating importance measures. Results from Step 3, as discussed in Section 2.3.5, are either "No change", "Elevate one level", or "Reduce one level ." The final column of Table 2.2 gives the results from this Step 4 evaluation. The recommendations for HF review are given as Level I, II, or III.

Table 2.2 Integrated assessment with RI screening

Results of Step 1 RG 1.174 (see Note 1)	Results of Step 2 Importance Measures	Results of Step 3 Qualitative Assessment	Results of Step 4 Recommended Level of HF Review
Region I (HA only)	-	-	Level I
Region I (Equipment & HA)	Level I	No change or elevate	Level I
		Reduce	Level II
	Level II	Elevate	Level I
		No change	Level II
		Reduce	Level III
	Level III	Elevate	Level II (see Note 2)
		No change or Reduce	Level III
Region II	Level I	No change or Elevate	Level I
		Reduce	Level II
	Level II	Elevate	Level I
		No change	Level II
		Reduce	Level III
	Level III	Elevate	Level II
		No change or Reduce	Level III
Region III	Level I	No change or Elevate	Level I (see Note 3)
		Reduce	Level II
	Level II	Elevate	Level I (see Note 3)
		No change	Level II
		Reduce	Level III
	Level III	Elevate	Level II
		No change Or Reduce	Level III

Notes:
1. If the modification being reviewed is a temporary one, then the entries of the first Column of this table will be Region I^{intgr}, II^{intgr}, & III^{intgr}.
2. This a Region I modification where the HA was determined not to be risk significant. However, qualitative factors suggested there was sufficient cause to recommend elevating the level of HF review. In such a situation, the analyst may determine that a further increase to a Level I review is justified, for example, where the qualitative analysis resulted in many "yes" responses or where the action is associated with a critical safety function.
3. This a Region III modification where the HA was found to be risk significant. An example of how this may occur is in a BWR with a modification to the RHR Suppression Pool Cooling to improve its reliability. This would then decrease risk, placing the overall modification in Region III. But in most BWRs, the operator action of SPC, has a very high RAW value (giving it a Level I from Step 2). Such a modification may or may not involve changes to the operator action. The final level of review will be based heavily on judgement, using the qualitative factors. As an example, Step 3 may have led to a recommendation to keep the review Level I or to increase it from Level II to Level I. The analyst then may determine that a Level I review is not warranted and prefer instead to conduct a Level II review. This may occur where the HA is simple, well defined, it is not time-pressured, and it is well supported by training, procedures, and the HSI design.

2.4 Screening Process for Non-risk-informed Change Requests

If the licensee's change request is not risk-informed, then a different process than that described in Section 2.3 is used. The safety significance of the HAs involved in the change request still is determined, but without benefit of the risk inputs that are provided by the licensee in a RI request. In this case there is no PRA information submitted by the licensee and the NRC staff performs a scoping type risk evaluation to estimate the risk importance of the change to the HA.

2.4.1 Overall Screening Approach for Non-risk-informed Requests

The overall screening approach for non-RI requests parallels that for RI requests, with two modifications. First, the reviewer verifies whether a non-RI request is appropriate, following the general guidance of SRP, Chapter 19. Second, the safety significance of HAs involved is determined using general risk and human reliability concepts, rather plant-specific risk information.

The non-RI screening process consists of the following steps:

1. Verifying that a non-RI change request is appropriate,
2. Assessing safety significance of the HA by either of two methods (Section 2.4.2 and 2.4.3),
3. Qualitatively assessing the safety significance of an HA involved in the request for change (Section 2.4.4), and
4. Making an integrated assessment of HA safety significance for determining the appropriate level of HF review (i.e., Level I, 2, or 3) (Section 2.4.5).

2.4.2 Assessment of Appropriateness of Non-risk-informed Submittal with Human Actions

Appendix D in Chapter 19 of the Standard Review Plan addresses the use of risk information in reviewing requests non-risk-informed license amendments. This assessment is performed by a risk analyst. In particular, the guidance given in SRP, Chapter 19 identifies:

1. When the risk implications of a non-risk-informed submittal would be discussed with a risk analyst, and

2. Examples of the potential impacts of "special circumstances" for which NRC staff can request risk information from the licensee, or reject the application.

Because non-RI licensee change requests may contain only HAs, the SRP, Chapter 19 guidance on these two topics is briefly discussed here. The discussion is not intended to replace that given in SRP, Chapter 19. Rather, it is intended to make reviewers aware of the guidance available there.

If the licensee decides to provide risk information to support the change request, as a result of the guidance immediately below, then the screening process for risk-informed applications is used. If not, then the following screening process for non-RI applications is employed.

2.4.2.1 Risk Implications of a Non-risk-informed Submittal

The SRP, Chapter 19 guidance, on when to discuss a submittal with a risk analyst, is duplicated here. As stated in Appendix D, "...the risk implications of a non-risk-informed submittal would be discussed with a risk analyst if the submittal -

- Significantly changes the allowed outage time (e.g., outside the range previously approved at similar plants), the probability of the initiating event, the probability of successful mitigative action, the functional recovery time, or the operator action requirement;[3]

- Significantly changes functional requirements or redundancy;

- Significantly changes operations that affect the likelihoods of undiscovered failures;

- Significantly affects the basis for successful safety function; or

- Could create 'special circumstances' under which compliance with existing regulations may not produce the intended or expected level of safety and plant operation may pose an undue risk to public health and safety."

In this guidance, "operator action" is explicitly stated in one of the above criteria. However, human actions can be involved in many other changes listed above (e.g., probability of successful mitigative action, functional recovery time, likelihood of undiscovered failures). If the licensee change request is judged to involve any of the above concerns, then, as stated in SRP, Chapter 19, "...[the licensee request] would be referred for a more detailed risk evaluation as part of the license amendment review."

2.4.2.2 Potential Effects of "Special Circumstances"

In this guidance, the description of "special circumstances" and examples of their potential impacts given in SRP, Chapter 19 are repeated. It is intended to assist reviewers of non-RI licencee change requests that involve HAs.

First, the SRP, Chapter 19 states that in general, a special circumstance may exist if

- The situation was not identified or specifically addressed in developing the current set of regulations and could be important enough to warrant a new regulation (e.g., a risk-informed regulation) if such situations were widespread, and

- The reviewer has knowledge that the risk impact is not reflected by the analysis of the licensing basis and believes that the risk increase would warrant denial, or attaching conditions to the staff's approval if the request were evaluated as a risk-informed application.

The SRP, Chapter 19 further gives the following examples of the potential impacts of "special circumstances." Additionally, some underlined phrases suggest interpretations for how HAs might be part of "special circumstances." If approved, licensee change requests that involve "special circumstances" could

---

[3] The term "requirements" as used here and elsewhere in this document, refers to requirements that are established as part of the design process. The term requirements is used in this context as a term-of-art. These are not "regulatory" requirements. There are no regulatory requirements in this document, only review guidance.

- Substantially increase the likelihood or consequences of accidents that are risk significant but are beyond the plant's design and licensing basis. For example, proposed changes to the steam generator (SG) allowable leak rates that meet 10 CFR Part 100 limits based on the design basis source term, but result in a large early release given a severe accident source term; or use of new materials for SG repairs that perform acceptably under normal and design basis conditions but have a reduced capability to maintain SG tube integrity in high-temperature, severe accident scenarios.

- Degraded multiple levels of defense or cornerstones in the reactor oversight process, through plant operations (including operator actions) or situations not explicitly considered in developing the regulations. For example, advanced applications of digital instrumentation and controls in which the licensee does not address, or comply with, regulatory guidance on evaluating the defense in depth and diversity in such digital systems.

- Significantly reduce the availability or reliability of structures, systems, components, or human actions that are risk significant but are not required by regulations. For example, amendment requests that, as an unintended consequence, compromise the effectiveness of the Mark I hardened wetwell vent system in protecting against containment overpressure failures in accidents beyond the design basis, or the diversity of the turbine-driven auxiliary feedwater pumps installed in response to NUREG-0737, Section II.E.1.1.

- Involve changes for which the synergistic or cumulative effects could significantly impact risk. For example, requests to uprate power that would increase operating power well beyond the levels approved in previous uprates, and would introduce or substantially increase the frequency of risk-significant core damage sequences.

Beyond these examples, the SRP, Chapter 19 states that if the reviewers "...believe that approval of the licensee change request would compromise the safety principles described in Regulatory Guide 1.174 and substantially increase risk relative to the risk acceptance guidelines contained in the regulatory guide, the reviewers should inform NRC management of the risk concerns and the need to further evaluate the risk associated with the request." The full guidance in the SRP should be consulted when reviewers believe that there might be such a concern associated with a non-RI submittal.

2.4.3 Assessment of Human Action Safety Significance

Two different methods are discussed for determining the safety significance of HAs. The first method requires a risk analyst to estimate the risk importance of the HA. The second, more generic method, is available if resources do not include a PRA analyst. This second method can be performed by the HF analyst. Either Method 1 or Method 2 can be used. Outputs from this step are then used in Section 2.4.5, Integrated Assessment of HA Significance for Non-RI Requests.

2.4.3.1 Method 1: Estimated Safety Significance of Human Action

In this method, a PRA analyst makes an assessment, using general PRA knowledge together with an understanding of the specific plant design. The result of this assessment is a surrogate for the RAW importance measure for the HA.

The following basic steps are used in making this assessment:

- Based on the definition of the HA, identify the impact of its failure on the key safety function(s) it supports:

    - Would this result in a loss of redundancy for a single system supporting a key safety function?
    - Would it result in the loss of a complete system?
    - Would it result in total failure of a key safety function?

- Determine the remaining capability to perform the key safety function, given the failure of the HA.

- Identify the accident sequences affected by the HA's failure. For most PRA modeling styles, this identification should be relatively straightforward based on understanding the event trees at the functional or systemic level, and the relationship between systems and key safety functions.

- Assess the significance of the failure of the HA based on an understanding of how much margin was eroded by its failure.

Using the steps above, together with generic values for initiating event frequencies, and reasonable estimates for system unavailabilities, a PRA analyst can roughly estimate the RAW risk importance measure for the HA. Once such RAW risk importance estimates have been generated, then Figure 2.5, RAW vs. New Baseline CDF, can be used to determine the initial estimate for the level of HF review. This level is entered in column 1 of Table 2.4.

2.4.3.2 Method 2: Generic Safety Significance of Human Action

When a PRA analyst is not available to undertake the approach described above, a more generic and less quantitative method can be used by the HF analyst to evaluate the safety significance of the HA.

The generic method for determining HA safety significance and subsequent level of human factors review, is based upon general risk information and some plant-specific information, whenever possible. However, the analyst should be cautioned that this generic approach is limited and may not always be conservative. For example, generic HAs that are risk-important are not necessarily representative of plant-specific risk importances. In addition, the change request may portray different performance conditions than those delineated by generic HA definitions, or even previously submitted plant-specific HA definitions. Finally, any new HAs (i.e., HAs not previously modeled generically or in a plant-specific PRA) may never have been analyzed and may be highly safety significant, regardless of the significance of the involved system or function in generic or plant-specific PRA results that do not reflect the requested change.

To determine HA safety significance with the generic approach, the HA involved in the change request is compared with the appropriate list of HAs for BWRs and PWR , given in Tables A.1 and A.2, respectively. These HAs were initially identified and grouped based on the risk-informed assessment process (Azarm, Higgins, and Chu, 1999), and from NUREG-1560. The grouping was then updated in 2001/2002 based on risk information from the latest licensee PRAs obtained during the site benchmarking of the Risk-informed Inspection Notebooks. These were developed specifically as slightly conservative evaluation tools.

The HAs in Tables A.1 and A.2 are organized into two groups. Group 1 contains the most risk-important HAs. RAW calculations typically would place them in the Level I area of Figure 2.5. Group 2 HAs are considered to be "potentially" risk-important. That is, they would be screened into the Level I review category for some plants, but not all. Typically, they affect risk, but not as significantly as the Group 1 actions. However, at some plants they may be quite risk-important. They were included in the second section of the plant Risk Informed Matrices (RIMs) as potentially important HAs.

Table 2.3 summarizes the simple and conservative logic on how to use the Group 1 and Group 2 HA assignments. HAs assigned to Group 1 are placed in the Level I review category. Those assigned to Group 2 go in either the Level I or Level II review category. If no risk submittal is made and the plant modification involves more than a minor change to a Group 2 action, then the NRC reviewer needs to decide whether it merits a Level I or Level II review. The conservative approach is to preliminarily assign it to Level I and then proceed to the Step 3, Qualitative Evaluation. If the reviewers have additional information about the risk of the action, either from the licensee or from the NRC's risk staff, they may choose to preliminarily assign the Group 2 HA to Risk Region II, and then proceed to the Step 3, Qualitative Evaluation.

It is important to note that, on a plant-specific basis, actions *not* listed in Tables A.1 and A.2 *may* also be risk-significant, and thus could require either a Level I or II review. This is not expected to be common but could happen. Therefore, if an action is not listed on either table, it cannot be concluded that it is not important to risk.

Thus, if no risk submittal is made and the plant modification involves an action that is not in Group 1 or 2, then an additional step is taken to determine whether it involves risk-important systems for the plant. This step can be used for both new and modified HAs. The risk-important systems can be obtained from the plant's IPE or latest updated PSA. This information also can be extracted from the plant-specific, risk-informed, inspection notebooks and related benchmarking reports that were completed by the NRC. For example, systems that benchmark as Red items should be considered to have high risk importance. Those that benchmark as Yellow or White should be considered as of moderate risk importance, while Green ones would have lower risk importance.

All plants now have a plant-specific, risk-informed, inspection notebook based on a recent updated version of the plant's PRA and a related benchmarking report that compared the risk importance (or coloring) of all of the major components and human actions in the notebook to the PRA. These are available to the NRC and to licensees. The risk importance of systems, components, or human actions can be determined by using the notebook per SECY-99-007A (NRC, 1999) and NRC IMC 0609 (NRC, 2003). These importances also can be obtained from Table 1, Summary of Benchmarking Results, contained in each of the plant-specific benchmarking reports. Assistance in using these documents, can be obtained by contacting NRC Senior Reactor Analysts (SRAs).

If the action in a non-RI submittal involves a high risk importance system (per the above paragraphs), and more than minor changes are involved, then the HA is considered most likely in risk Region I or II of RG 1.174. The same logic as discussed for Group II HAs applies, and the reviewer should preliminarily select a Level I or II review and then proceed to Step 3, Qualitative Evaluation. Similarly, if the HA involves a system of moderate importance, the HA should be considered in Region II of RG 1.174. If the modification involves only systems with lower risk-importance, it is initially considered to require a Level III review.

The logic applied to HAs in the generic method is conservative and may place an HA in a higher risk region than it would receive using plant-specific RAW calculations. Thus, it is beneficial for both licensees and NRC staff to use plant-specific risk information to more properly allocate review resources.

Note that the last two rows of Group 2 in Tables A.1 and A.2 are very general and written so that they will identify any new HAs that did not exist previously and are most likely not in the PRA, but which may be risk important.

The principal motivation for these assignments is to make conservative assessments. Further adjustments can be made in Step 4 (i.e., the integrated assessment) with inputs from the qualitative assessment.

Column 3 of Table 2.3 below provides the initial level assignment based on this subsection 2.4.3.2. This should be entered into the first column of Table 2.4. The review then proceeds to the Qualitative Assessment of Section 2.4.4 below.

Table 2.3 Generic approach for placing HAs into HF review levels

Generic Groups that contain the HA	Systems involving the HA	Level of HF Review for the HA
Group 1	NA	Level I
Group 2	NA	Level I or II*
Neither Group	High risk importance (Red)	Level I or II*
Neither Group	Moderate risk importance (Yellow or White)	Level II
Neither Group	Low risk importance (Green)	Level III

*See discussion in text of Section 2.4.3.2 for determination of Level I or II here.

2.4.4 Qualitative Assessment of Human Action Safety Significance

The qualitative assessment of HAs involved in non-RI change requests is identical to that for RI change requests. Consequently, the guidance in Section 2.3.5 is appropriate and should be used here. Outputs from this step are used in Section 2.4.5 below, "Integrated Assessment for HA Safety significance."

2.4.5 Integrated Assessment of Human Action Safety Significance for Non-risk-informed Requests

The integrated assessment of HA safety significance for non-RI applications is similar to that for RI applications, but simpler because there are fewer inputs to integrate. This integrated assessment is performed by the HF analyst. Recommendations for the level of HF review using the non-RI screening process are shown in Table 2.4.

The basis for the recommended HF review levels shown in Table 2.4 are similar to those shown in Table 2.2 for the RI screening process. However, because the HA importance estimates produced in Step 1 of the non-RI screening process are expected to be much less sophisticated than those for the RI screening process, the qualitative assessment results may be given more weight in the non-RI screening process. The goal in these recommendations, as for those in the RI screening process, is to produce reasonable but conservative assessments. In line with this goal, consider the Table entries for a Level I or Level II HA, where the Qualitative Assessment Results recommend "Reduce." The Table maintains the option to not reduce the level of review as part of this Section 2.4.5, Integrated Assessment. This is done to add caution in reducing the level of review for a HA that was evaluated using the generic method.

Table 2.4 Integrated assessment with non-RI screening

Results of Section 2.4.3 HA Safety Significance	Results of Section 2.4.4 Qualitative Assessment Results	Results of Section 2.4.5 Integrated Assessment
Level I	No change OR Elevate	Level I
	Reduce	Level I or II
Level II	Elevate	Level I
	No change	Level II
	Reduce	Level II or III
Level III	Elevate	Level II*
	No change OR Reduce	Level III

* This a modification where the HA was determined not to be risk significant based on generic analyses only. However, based on qualitative factors, there was sufficient cause to recommend elevating the level of HF review. In such a situation, the analyst may determine that a further increase to a Level I review is justified, for example, where the qualitative analysis resulted in many "yes" responses and where there is uncertainty about the importance of the action.

2.5   Level of Human Factors Review for Human Actions

Once the appropriate screening process for either RI or non-RI requests is performed, the level of human factors engineering review to be performed is determined in section 2.4 above. Three levels of review are provided with the most important HAs (i.e., Level I HAs) receiving the most thorough review. Less important HAs (Level II) will receive a more efficient review that is appropriate to their level of importance. HAs that are assigned to Level III will receive no (or minimal) review. The principal focus of a Level III HA review is to verify that these HAs have been properly classified in Level III using the screening process.

Briefly, the three levels of HF reviews are:

Level I   This level corresponds with the most extensive and detailed HF review, as described in Section 3. This level includes review of planning, analyses, design, and verification and validation activities, and a performance monitoring strategy.

Level II   This level corresponds with a less detailed review. Section 4 provides guidance for performing a Level II review. In special circumstances, as identified in Step 3, NRC may choose to add selected Level I review criteria to a Level II review, rather than elevating the HA to a full Level I review.

Level III   This level corresponds with a minimal review. The minimum aspects of this review include:

- Licensee documentation and NRC verification of the appropriateness of the Level III assignment for the HA(s) associated with the change request

- NRC verification that current regulations are still being met with the change in place (using Criterion 1 in Section 3.1, "General Deterministic Review Criteria ").

In addition, licensees are encouraged to utilize the Level II guidance contained in Section 4 to ensure that the HAs can be accomplished as assumed. Also, in some cases, there may be one or two aspects, identified during the Step 3 qualitative assessment, that the NRC chooses to review. NRC would use the appropriate portions of the Level II review guidance for these identified aspects.

3   LEVEL I REVIEW GUIDANCE

The guidance in this section is a tailoring of NUREG-0711, (NRC, 2004b) to plant modifications affecting HAs of high risk-importance. NUREG-0711, Section 1.4, Graded Approach to Review, indicates that the level of staff review of an applicant's human factors engineering (HFE) design should reflect the unique circumstances of the review and that the guidance should be selectively applied to address the demands of each specific review.

The tailoring was accomplished by selecting the NUREG-0711 criteria that were appropriate to the review of Level I HAs that are risk-important and then modifying these criteria to better reflect the context of the types of plant modifications involved. This guidance has been developed to "stand alone." That is, aspects of the review criteria that were not changed are repeated in this section rather than referring the reviewer back to NUREG-0711. This makes the guidance easier to use.

Even with the tailored guidance provided in this section, the reviewer can further adapt the guidance to meet the unique demands of particular reviews.

3.1   General Deterministic Review Criteria

Objective

The objective of this review is to verify that deterministic aspects of design, as discussed in RG 1.174, have been appropriately considered by the licensee. Deterministic aspects include: ensuring the change meets current regulations, and does not compromise defense-in-depth.

Scope

The deterministic review criteria apply to all modifications associated with Level I HAs.

Criteria

(1)   The licensee should provide adequate assurance that the change meets current regulations, except where specific exemptions are requested under 10 CFR 50.12 or 10 CFR 2.802. For example, a change might be identified as risk significant when using a standard PRA to screen for risk. However, an exemption might be granted under one or more of the following regulations: 10 CFR 20, 10 CFR 50 Appendix A, Criterion 19, and10 CFR 50 Appendices C through R.

(2)   The licensee should provide adequate assurance that the change does not compromise defense-in-depth. Defense-in-depth is one of the fundamental principles upon which the plant was designed and built. Defense-in-depth uses multiple means to accomplish safety functions and to prevent the release of radioactive materials. Defense-in-depth is important in accounting for uncertainties in equipment and human performance, and for ensuring some protection remains even in the face of significant breakdowns in particular areas. Defense-in-depth may be changed but overall should be maintained. Important aspects of defense-in-depth include:

- A reasonable balance is preserved among prevention of core damage, prevention of containment failure, and consequence mitigation.

- There is no over-reliance on programmatic activities to compensate for weaknesses in plant design. This may be pertinent to changes in credited operator actions.

- System redundancy, independence, and diversity are preserved commensurate with the expected frequency, consequences of challenges to the system, and uncertainties (e.g., no risk outliers).

- Defenses against potential common cause failures are preserved, and the potential for the introduction of new common cause failure mechanisms is assessed. Caution should be exercised in crediting new operator actions to provide adequate assurance that the possibility of significant common cause operator errors are not created.

- Independence of barriers is not degraded.

- Defenses against human errors are preserved. One way to help ensure this for HAs that are risk-important is to establish procedures for a second check or independent verification that such important actions have been properly executed.

- Safety margins often used in deterministic analyses to account for uncertainty and provide an added margin to provide adequate assurance that the various limits or criteria important to safety are not violated. Such safety margins are typically not related to HAs, but the reviewer should take note to see if there are any that may apply to the particular case under review. It is also possible to add a safety margin (if desired) to the HA by requiring a demonstration that the action can be performed within some time interval (or margin) that is less than the time required by the analysis.

3.2   Operating Experience Review

Objective

The objective of this review is to verify that the licensee has identified and analyzed HFE-related problems and issues encountered previously in designs and human tasks that are similar to the planned modification so that issues that could potentially hinder human performance can be addressed.

Scope

The operating experience review (OER) encompasses all proposed changes to HAs and addresses the operating histories of plant systems, HAs, procedures, and HSI technologies. The scope of the HSI technology review can be graded as follows:

(1)   If existing HSI components are to be used without modification and if they are currently used for safety-related functions within the plant, then a review of the operating experience with those HSI components is not necessary.

(2)   If existing HSI components are to be used without modification but they are not currently used for safety-related functions then the operating experience with those HSI components should be reviewed.

32

(3)   If new HSI components are to be installed or the existing HSI is to be modified using HSI technologies that have not been previously used in the plant for safety-related functions then the operating experience with those HSI components should be reviewed.

Criteria

(1)   *Plant Systems* - The licensee's review should include information pertaining to the operation and maintenance of the plant system prior to the change in the HAs.

(2)   *Human Actions* - The licensee's review should identify performance issues associated with procedural guidance, training, and HAs for the system prior to the proposed change to the actions, including the types of actions performed, the procedures available for those actions, and the adequacy of those procedures.

(3)   *HSI Technologies* - The licensee's review should identify human performance issues associated with HSI technologies for the proposed changes in the HAs, if they are different from those used successfully at their plant.

(4)   *Issues Identified by Plant Personnel* -Interviews and surveys with personnel should be conducted to determine operating experience related to the plant system before the change in the HAs. Discussions of plant operations and HFE/HSI design should be limited to topics relevant to the change in the HAs.

(5)   *Use for Design Input* - Issues identified by the OER should be used for input to the design of modifications to the HSI, procedures, and training, and tracked to provide assurance that they are addressed.

3.3   Functional Requirements Analysis and Function Allocation

(Note: If there are no changes in Functional Requirements or Functional Allocation from the current plant design, this review element is not needed.)

Objective

The objective of this review is to verify that the licensee has:

(1)   Defined any changes in the plant's safety functions (functional requirements analysis), and

(2)   Provided evidence that the allocation of functions between humans and automatic systems provides an acceptable role for plant personnel; i.e., the allocations take advantage of human strengths and avoid functions that would be negatively affected by human limitations (functional allocation).

Scope

This review addresses all plant functions affected by the change in operator actions including changes to the functions and to their allocation between personnel and automatic systems. The level of detail in the functional requirements and allocation analyses may be graded by the reviewer based on: (1) the degree of difference between the HAs before and after the change; (2) the extent to which difficulties occurred in prior operations, as identified through the OER; and (3) the risk level associated with the change. The following additional considerations apply:

(1)   If new safety functions are introduced or existing ones changed, then reviews of both the functional requirements analysis and function allocation analysis should be conducted. (This situation is not likely to occur since it would involve a significant deviation from the design basis that was originally approved by the NRC.)

(2)   If the function allocation is changed, or if the risk level is well into Level I (as determined by the PRA/HRA review criteria) then a review of the function allocation should be conducted. (Many cases will have changed function allocations. An example may be the reallocation of responsibility from an automatic system to personnel for the initiation, on-going control, or termination of a function.)

(3)   If the function allocation is not changed then no function allocation analysis is needed and the licensee should proceed with task analysis. (An example may be a manual action performed for a safety-related function that is now required under a new scenario. That is, the function is the same but the initiating circumstances are different.)

Review Criteria

Functional Requirements

(1)   New or changed safety functions should be described, including comparisons before and after the proposed change. The set of plant system configurations or success paths that are responsible for or capable of carrying out the safety function should be clearly defined and the ones affected by the proposed changes in the HAs should be identified. This functional decomposition should address:

- High-level functions [e.g., maintain reactor coolant system (RCS) integrity] and critical safety functions (e.g., maintain RCS pressure control)

- Specific plant systems and components

- Technical basis for changes to functions

Functional Allocation

(1)   For the functional allocation analysis, a description should be provided for each of the high-level functions allocated to the human as a result of the proposed change. The description should include the following:

- Purpose of the high-level function

- Conditions under which the high-level function is required

- Parameters that indicate that the high-level function is available

- Parameters that indicate the high-level function is operating (e.g., flow indication)

34

- Parameters that indicate the high-level function is achieving its purpose (e.g., reactor vessel level returning to normal)

- Parameters that indicate that operation of the high-level function can or should be terminated

Note that parameters may be described qualitatively (e.g., high or low), rather than as specific numerical values or setpoints.

(2)   The technical basis for all relevant functional allocations should be documented. The basis for function allocations can be successful operating experience. This analysis should reflect (a) sensitivity, precision, time, and safety-related requirements; (b) required reliability; and (c) the number and level of skills of personnel required to operate and maintain the system.

(3)   The allocation analysis should consider not only the personnel role of initiating manual actions but also responsibilities concerning automatic functions, including monitoring the status of automatic functions to detect system failures. The demands associated with the proposed allocation of functions should be considered in terms of all other human functions that may impose concurrent demands upon the personnel. The overall level of workload should be considered when allocating functions to the personnel.

3.4   Task Analysis

Objective

The objective of this review is to verify that the licensee's task analysis (TA) identifies the behavioral requirements of the tasks personnel are required to perform. The task analysis should form the basis for specifying the requirements for the HSI, procedures, and training based on the tasks personnel will perform. The results are also used as basic information for developing staffing and communication requirements of the plant.

For a change to an existing action, a new TA may not be necessary.

Scope

The task analysis addresses HAs in their entirety, including all pertinent plant conditions, situational factors, and performance shaping factors.

Criteria

(1)   The licensee should identify the information that is required to inform personnel that each HA is necessary, that the HA has been correctly performed, and that the HA can be terminated.

(2)   Plant personnel who are affected by the HAs should be identified, including licensed control room operators as defined in 10 CFR Part 55 and the following categories of personnel defined by 10 CFR 50.120: nonlicensed operators, shift supervisor, shift technical advisor, instrument and control technician, electrical maintenance personnel, mechanical maintenance personnel, radiological protection technician, chemistry technician, and engineering support personnel.

(3)   Task analyses should provide detailed descriptions of what the personnel must do. The licensee should identify how human tasks or performance requirements are being changed.

(4)   The task analysis should address the full range of plant conditions and situational factors, and performance shaping factors anticipated to influence human performance. The range of plant operating modes relevant to the HAs (e.g., abnormal and emergency operations, transient conditions, and low-power and shutdown conditions) should be included in the task analysis.

(5)   The human task requirements that result from the changes in the actions should be assessed to determine whether they are compatible with each individual's responsibilities (i.e., will not interfere with or be disrupted by the cognitive and physical demands of other tasks and responsibilities).

(6)   The task analysis should identify reasonable or credible, potential errors.

3.5   Staffing

Objective

The objective of this review is to verify that the licensee has analyzed the proposed change in HAs to determine the number and qualifications of personnel based on task requirements and applicable regulatory requirements. Adding additional manual actions or shifting tasks to periods of high workload may increase staffing needs. An example is a local manual action to temporarily replace an automatic action.

Scope

The staffing analysis addresses personnel needs for all conditions in which the HA may be performed.

Criteria

(1)   Staffing levels should be evaluated to determine their adequacy with respect to any additional burden that may be imposed by the plant or HA modifications. The staffing levels should be adjusted if necessary. The evaluation should be based on an analysis of:

- Current nominal (typical shift complement of personnel) and minimal staffing levels (as identified in administrative procedures)

- Required actions determined from the task analysis, if performed

- The physical configuration of the work environment (e.g., control room and control consoles configurations that may affect the ability of personnel to work together)

- The availability of plant information from individual workstations from individual and group view components of the HSI

- Availability of personnel considering other activities that may be ongoing and for other possible responsibilities outside the control room (e.g., fire brigade)

3.6   Probabilistic Risk and Human Reliability Analysis

Objective

The objectives of this review are to verify that the licensee has: (1) updated the PRA model to reflect system, component, and HA changes that may be necessary based on the proposed modification or HAs; (2) performed an analysis of the potential effects of the proposed changes on plant safety and reliability, in a manner consistent with current, accepted PRA/HRA principles and practices, and (3) verified that the risk insights derived from the results are addressed in the selection of HAs; development of procedures, HSI components, and training in order to limit risk and the likelihood of personnel error and to provide for error detection and recovery capability.

Scope

This review addresses PRAs and HRAs conducted by the licensee to evaluate changes in systems, components, and human tasks that result from the proposed changes in HAs. Some of these items may have been addressed and reviewed as part of the screening process for the risk-informed submittal. In addition, the NRC human factors engineering reviewers may consult with the NRC risk analysis specialist on this review.

Criteria

(1)   The PRA and HRA should be modified to reflect the changes in systems, components, and human tasks. Human interactions with plant systems and components should be analyzed at least at the level modeled in the plant's current PRA. Alternatively, justification should be provided as to why the change to the PRA is minimal and not necessary.

(2)   The HRA should follow a structured, systematic, and auditable process to provide adequate assurance that the reliability of each HA is accurately estimated so that its effect on plant safety using the PRA can be assessed.

(3)   The PRA/HRA should address any human interactions that may be involved with the modified plant systems and components at the level currently modeled in the plant PRA, for example,

- Errors of omission and commission

- Miscalibration and component restoration errors

- Recovery actions

(4)   The analysis of HAs should include the identification of performance shaping factors (PSFs), that is, factors that influence human reliability through their effects on performance. PSFs include factors such as environmental conditions, HSI design, procedures, training, and supervision.

(5)   HAs, associated with the modification, that are risk-important should be identified from the PRA/HRA and used as input to the design of procedures, HSI components, and training. These actions should be developed from the Level 1 (core damage) PRA and Level 2 (release from containment) PRA including both internal events and external events (if available). They should be developed using selected (more than one) importance measures and HRA sensitivity analyses

37

to provide adequate assurance that an important action is not overlooked because of the selection of the measure or the use of a particular assumption in the analysis.

(6)   The licensee should use the information from the modified PRA/HRA to calculate changes in CDF, LERF, and integrated risk (if a temporary change is involved).

3.7   Human-system Interface Design

Objective

The objective of this review is to evaluate the HSI design, for those changes in HAs that require changes to the HSI, to verify that the licensee has appropriately translated function and task requirements into the detailed design of the HSI through the systematic application of HFE principles and criteria.

Scope

This review addresses the design of temporary and permanent modifications to the HSI, including new HSI components and the modification of existing ones, related to the proposed changes in the HAs, to verify that the existing HSI are appropriate for the modified human action.   The review addresses aspects of the HSI and the work environment that affect the ability of the personnel to perform the HAs.

Criteria

(1)   The HSI should be designed consistent with HFE guidelines and the existing HSI to the degree practical.

(2)   The design should seek to minimize the probability that errors will occur and maximize the probability that errors will be detected and personnel will be able to recovered from them.

(3)   When developing HSI components for actions performed either in the control room or locally in the plant, the following factors should be considered:

- Communication, coordination, and workload

- Feedback

- Local environment

- Inspection, test, and maintenance

(4)   The layout of HSI components within consoles, panels, and workstations should be based upon (1) analyses of human roles (job analysis) and (2) systematic strategies for organization such as arrangement by importance, frequency of use, and sequence of use.

(5)   HSI characteristics for the changed action should support human performance under the full range of environmental conditions.

(6)   Certain human tasks will need qualified instrumentation in accordance with RG 1.97 (NRC, 1983). The task analysis should identify the necessary safety grade of the control and display equipment used for human tasks. The RG defines Type A variables as "those variables to be monitored that provide the primary information required to permit the control room operators to take the specified manually controlled actions for which no automatic control is provided and that are required for safety systems to accomplish their safety function for design basis accident events" (NRC, 1983, p. 1.87-4). Primary information is further defined in the RG as information that is essential for the direct accomplishment of the specified safety functions, but does not include those variables that are associated with contingency actions that may also be identified in written procedures. Table 1 of RG 1.97 provides detailed Category 1 criteria that Type A variables should meet. In general, these Category 1 criteria provide for environmental and seismic qualification, redundancy, quality assurance, continuous display, good human factors design, and an emergency power supply. Therefore, HAs, which are required for safety systems to accomplish their safety function for design basis accident events and for which no automatic control is provided, will need control and display instrumentation in accordance with RG 1.97. (This RG allows for consideration of alternative approaches that are adequately justified and include consideration of the risk significance of the actions involved.) Thus, credit should only be given for these types of HAs if they can be completed using control and display instrumentation that is consistent with RG 1.97. Information in RIS 2002-22 (NRC 2002a) may also be useful.

3.8   Procedure Design

Objective

The objective of this review is to verify that applicable plant procedures have been appropriately modified, where needed, to provide adequate guidance for the successful completion of the HAs, and that the procedures adequately reflect changes in plant equipment and HAs. In the procedure development process, HFE principles and criteria should be applied along with all other design requirements to develop procedure modifications that are technically accurate, comprehensive, explicit, easy to use, and validated.

Scope

This review addresses all plant procedures that provide guidance to personnel for the affected actions, including the following types:

•   Emergency operating procedures (EOPs)

•   Plant and system operations (including startup, power, and shutdown operations)

•   Abnormal and emergency operations

•   Alarm response

The scope includes both temporary and permanent modifications to these procedures.

Criteria

(1)   Where applicable, plant procedures should be modified to provide new instructions for the proposed changes in the HAs. Exceptions may be made where the adequacy of the existing procedures can be justified. Such a justification should indicate how the existing procedures provide necessary and sufficient guidance for the changed HAs and do not contain information that is inaccurate or no longer relevant.

(2)   Where appropriate, procedures should identify how the operating crew should independently verify that the HAs have been successfully performed.

(3)   All procedures should be verified and validated to provide adequate assurance that they are correct and can be carried out. Their final validation should be performed as part of the validation activities described in Section 3.10.

(4)   Any changes in the HSI should be reflected in the modifications of the procedures.

(5)   Procedural modifications should be integrated across the full set of procedures; alterations in particular parts of the procedures should not conflict nor be inconsistent with other parts. For example, an HSI component that is modified for a HA may also affect other actions that have not been modified. Therefore, procedure changes should not be limited to only the changed HAs.

3.9   Training Program Design

Objective

The objective of this review is to verify that the licensee's training program results in adequate training for the HAs. The review should verify that appropriate training has been developed and conducted for the HAs, including any changes in qualifications, as described in NRC Information Notice 97-78 (NRC, 1997).

Scope

This review addresses the licensee's training programs for all licensed and non-licensed personnel who perform the changed HAs. The scope includes both temporary and permanent modifications to training programs.

Criteria

(1)   The licensee's training program should be modified to address the knowledge and skill requirements for all changes in HAs for the licensed and non-licensed personnel.

(2)   Learning objectives should be derived from an analysis that describes desired performance for the HAs after training has been completed.

3.10   Human Factors Verification and Validation

Objective

The objective of this review is to verify that the licensee's verification and validation (V&V) program:

- Provides adequate assurance that the HFE/HSI design contains all necessary alarms, displays, and controls to support plant personnel tasks (HSI Task Support Verification).

- Provides adequate assurance that the HFE/HSI design conforms to HFE principles, guidelines, and standards (HFE Design Verification).

- Provides adequate assurance that the HFE/HSI design can be effectively operated by personnel within all performance requirements applicable to the HAs, including the following Integrated System Validation):

- All pertinent staffing considerations are acceptable for nominal and minimal shift levels, such as shift staffing, assignment of tasks to crew members, and crew coordination within the control room and between the control room and local control stations and support centers.

- The HAs can be accomplished within time and performance criteria

- The integrated system performance is consistent with all functional requirements, including tolerance of failures of individual HSI features

Scope

(1)   The general scope of V&V includes the following factors as applicable to the proposed changes to the HAs:

- HSI hardware and software

- Procedures

- Workstation and console configurations

(2)   The typical order of V&V activities is:

- HSI task support verification

- HFE design verification

- Integrated system validation

(3)   All V&V activities are applicable regardless of whether the change in HAs involve changes in the HSI.

Criteria

<u>HSI Task Support Verification</u>

(1)   All aspects of the HSI (e.g., controls, displays, procedures, and data processing) that are required to accomplish the HAs should be verified as available through the HSI. For HAs associated with qualified instrumentation in accordance with RG 1.97, it should be verified that the HSI provides such qualified instrumentation.

<u>HFE Design Verification</u>

(1)   All aspects of the HSI (e.g., controls, displays, procedures, and data processing) used for the HAs should be verified as consistent with accepted HFE guidelines, standards, and principles.

(2)   Deviations from accepted HFE guidelines, standards, and principles should be acceptably justified on the basis of a documented rationale such as trade study results, literature-based evaluations, demonstrated operational experience, or tests and experiments.

<u>Integrated System Validation</u>

Validation Testbeds

(1)   For HAs performed in the main control room, the plant training simulator should be used as the testbed when conducting the validation tests.

(2)   For HAs performed at locations outside of the main control room, the use of a simulation or mockup can be used or drills conducted in the plant. The conduct of these drills should not interfere with plant operations (e.g., drills may be conducted when the plant is shutdown or the affected systems are removed from service).

Plant Personnel

(1)   Participants in the validation tests should be the plant personnel who will perform the changed actions. Actions that will be performed by licensed personnel should be validated using licensed personnel rather than training or engineering personnel. Similarly, actions allocated to non-licensed personnel should be validated using non-licensed personnel.

(2)   To properly account for human variability, more than one crew should participate in the validation tests. This will help provide adequate assurance that variation along most of the significant dimensions that influence human performance are included in the validation tests. Participation is not necessary for personnel who do not normally operate or maintain the plant (e.g., administrative personnel who hold operating licenses).

(3)   In selection of personnel, consideration should be given to the assembly of nominal and minimum crew configurations, including shift supervisors, reactor operators, shift technical advisors, etc., that will participate in the validation tests. The composition of operations personnel need only include categories of personnel that are relevant to the HAs.

Operational Conditions

(1)   Integrated system validation should consider the operational conditions for which each HA is required.

(2)   The operational conditions should be developed into scenarios. The following information should be defined to provide adequate assurance that important performance dimensions are addressed and to allow scenarios to be accurately presented for repeated trials:

- Description of the scenario mission and any pertinent "prior history"

- Specific initial conditions

- Events (e.g., failures) to occur and their initiating conditions, e.g., time, parameter values, or events

- Data to be collected and the specification of what, when and how data are to be obtained and stored

- Specific criteria for terminating the scenario

(4)   Scenarios should have appropriate task fidelity so that realistic task performance will be observed in the validation tests and so that results can be generalized to actual operation in the real plant.

(5)   When evaluating performance associated with the use of HSI components located remote from the main control room, the effects on crew performance due to potentially harsh environments (i.e., high radiation) should be simulated (i.e., additional time to don protective clothing and access radiologically controlled areas).

Plant Performance Measurement

(1)   The variables used in the performance measures should include performance of the plant and personnel, as described below.

(2)   Measures that assess personnel task performance should be used, including the following:

- For each specific scenario, the tasks that personnel *are required to* perform should be identified and assessed. Such tasks can include necessary primary (e.g., start a pump) as well as secondary (e.g., access the pump status display) tasks. This analysis should be used for the identification of errors of omission by identifying tasks which should be performed. The proper completion of required tasks should be verified.

- The tasks that are *actually* performed by personnel during simulated scenarios should be identified and quantified. The variable(s) used to quantify tasks should be chosen to reflect the important aspects of the task with respect to system performance, such as:

  - Task success or failure

  - Task completion time

43

- Errors (omission and commission)

- Subjective reports of participants

(3)   Performance criteria for the measures used in the evaluations should be established.

Data Analysis and Interpretation

(1)   Validation test data, time and errors, should be analyzed through a combination of quantitative and qualitative methods. For example, task time can be statistically compared with time available to perform the task and subjective reports of participants can be qualitatively evaluated to identify potential obstacles to performance.

3.11   Human Performance Monitoring Strategy

Objective

The objective of this review is to verify that the licensee has prepared a human performance monitoring strategy for ensuring that no adverse safety degradation occurs because of the changes that are made and to provide adequate assurance that the conclusions that have been drawn from the evaluation remain valid over time. A human performance monitoring strategy by the licensee will help to ensure that the confidence developed by the completion of the integrated system validation is maintained over time. There is no intent to periodically repeat the full integrated system validation, however, there should be sufficient evidence to provide reasonable confidence that operators have maintained the skills necessary to accomplish the assumed actions.

The results of the monitoring need not be reported to the NRC, but should be retained onsite for inspection.

Scope

The scope of the performance monitoring strategy should provide adequate assurance that the:

- HFE/HSI design can be effectively operated by personnel, both within the control room and between the control room and local control stations and support centers.

- HAs can be accomplished within time and performance criteria.

- Integrated system performance is maintained within the performance established by the integrated system validation.

Criteria

(1)   A human performance monitoring strategy should be developed and documented by the licensee. The strategy should be capable of tracking human performance after the changes have been implemented to demonstrate that performance is consistent with that assumed in the various analyses that were conducted to justify the change. Licensees may integrate, or coordinate, their performance monitoring for risk-informed changes with existing programs for monitoring operator performance, such as the licensed operator training program. If a plant change requires monitoring of actions that are not included in existing training programs, it may be advantageous

for a licensee to adjust the existing training program rather than to develop additional monitoring programs for risk-informed purposes.

(2) The program should be structured such that (1) HAs are monitored commensurate with their safety importance, (2) feedback of information and corrective actions are accomplished in a timely manner, and (3) degradation in performance can be detected and corrected before plant safety is compromised (e.g., by use of the plant simulator during periodic training exercises).

4   LEVEL II REVIEW GUIDANCE

The guidance in this section also is a tailoring NUREG-0711, (2004b) to plant modifications affecting HAs of medium risk significance. The guidance in this section reflects a further reduction of the criteria to reflect the level of risk imposed by the modification in Level II. Even with the tailored guidance provided in this section, the reviewer can further adapt the guidance to meet the unique demands of particular reviews.

4.1   General Deterministic Review Criteria

Objective

The objective of this section is to verify that deterministic aspects of design, as discussed in RG 1.174, have been appropriately considered by the licensee. Deterministic aspects include: verifying that the change meets current regulations; and does not compromise defense-in-depth.

Scope

The deterministic review criteria are applicable to all modifications associated with Level II HAs.

Criteria

(1)   The licensee should provide adequate assurance that the change meets current regulations, except where specific exemptions are requested under 10 CFR 50.12 or 10 CFR 2.802. Examples of regulations that may be affected by a change, but that may be identified as risk significant when using a standard PRA to screen for risk include the following: 10 CFR 20, 10 CFR 50 Appendix A, Criterion 19, and 10 CFR 50 Appendices C through R.

(2)   The licensee should provide adequate assurance that the change does not compromise defense-in-depth. Defense-in-depth is one of the fundamental principles upon which the plant was designed and built. Defense-in-depth uses multiple means to accomplish safety functions and to prevent the release of radioactive materials. It is important in accounting for uncertainties in equipment and human performance, and for ensuring some protection remains even in the face of significant breakdowns in particular areas. Defense-in-depth may be changed but overall should be maintained. Important aspects of defense-in-depth include:

- A reasonable balance is preserved among prevention of core damage, prevention of containment failure, and consequence mitigation.

- There is no over-reliance on programmatic activities to compensate for weaknesses in plant design.

- System redundancy, independence, and diversity are preserved commensurate with the expected frequency, consequences of challenges to the system, and uncertainties (e.g., no risk outliers).

- Defenses against potential common cause failures are preserved, and the potential for the introduction of new common cause failure mechanisms is assessed.

- Independence of barriers is not degraded.

- Defenses against human errors are preserved.

4.2   Analysis

Objective

The objective of the review is to verify that the licensee has analyzed the changes to HAs and identified HFE inputs for any modifications to the HSI, procedures, and training that may be necessary.

Scope

The review criteria are applicable to all modifications associated with Level II HAs.

Criteria

(1)   *Functional and Task Analysis*

- The licensee should identify how the personnel will know when the HA is necessary, that is performed correctly, and when it can be terminated.

- Task analyses should provide a description of what the personnel must do. The licensee should identify how human tasks or performance requirements are being changed. The task analysis should identify reasonable or credible, potential errors and their consequences.

(2)   *Staffing* - The effects of the changes in HAs upon the number and qualifications of current staffing levels of operations personnel for normal and minimal staffing conditions.

4.3   Design of Human System-interfaces, Procedures, and Training

Objective

The objective of the review is to verify that the licensee has supported the HAs by appropriate modifications to the HSI, procedures, and training.

Scope

The review criteria are applicable to all modifications associated with Level II HAs.

Criteria

(1)   *HSIs* - Temporary and permanent modifications to the HSI should be identified and described. The modifications should be based on task requirements, HFE guidelines, and resolution of any known operating experience issues.

(2)   *Procedures* - Temporary and permanent modifications to plant procedures should be identified and described. The modifications should be based on task requirements and resolution of any

known operating experience issues. Justification should be provided when the plant procedures are not modified for changes in operator tasks.

(3)   *Training* - Temporary and permanent modifications to the operator training program should be identified and described. The modifications should be based on task requirements and resolution of operating experience issues. Justification should be provided when the training program is not modified for changes in operator tasks.

4.4   Human Action Verification

Objective

The objective of this review is to verify that the licensee has demonstrated that the HAs can be successfully accomplished with the modified HSI, procedures, and training.

Scope

The review criteria are applicable to all modifications associated with Level II HAs.

Criteria

(1)   An evaluation should be conducted at the actual HSI to determine that all required HSI components, as identified by the task analysis, are available and accessible.

(2)   A walk-through of the HAs under realistic conditions should be performed to determine that:

- The procedures are complete, technically accurate, and usable

- The training program appropriately addressed the changes in plant systems and HAs

- The HAs can be completed within the time criterion for each scenario that is applicable to the HAs.

The scenario used should include any complicating factors that are expected to affect the crews ability to perform the HAs.

(3)   The walk-throughs should include at least one crew of actual operators.

5   FINAL DECISION ON ACCEPTANCE OF HUMAN ACTIONS

Once the NRC review of a proposed licensee change (either risk-informed or non-risk informed) in HAs is completed, a final decision regarding the acceptability of the human action aspect of the modification must be made. This decision will provide input to the NRC's overall decision whether to accept or reject the licensee's submittal. The final results of the human factors review will provide input to the Integrated Decisionmaking (see discussion in RG 1.174, Section 2.2.6) and may be documented in a Safety Evaluation Report provided to the NRC Project Manager.

At this point a significant amount of information has been gathered, reviewed, and evaluated that can be used to assist in the final decision-making. This information includes:

- The risk values related to the change or modification, including their location on the acceptance guideline figures

- The time associated with the change

- The results of the Level I or Level II review, which includes both human factors engineering information relating to the ability of operators to reliably perform the actions in question and deterministic aspects of the proposed change

- Answers to requests for additional information (RAIs) that NRC has developed providing additional information or commitments

- Other factors related to the plant in question that may bear on the decision

This information needs to be considered in an integrated fashion, that considers risk, but does not wholly base the final decision on risk alone. RG 1.174 notes that the use of PRA technology should be increased in all regulatory matters, but it should be done in a manner that complements the NRC's deterministic approach and supports the NRC's traditional defense-in-depth philosophy. RG 1.174 also notes that decisions concerning proposed changes are expected to be reached in an integrated fashion, considering traditional engineering and risk information, and may be based on qualitative factors as well as quantitative analyses and information. The review guidance in this document takes these concepts into consideration.

RG 1.174 notes that HAs in the high risk area of Region I are generally not desired, but there are examples of such actions in plants today, e.g., the PWR ECCS switchover situation described in Generic Issue B-17. Also, there may be extenuating circumstances in which the licensee can adequately justify a modification to add a Region I HA, e.g., if the change is temporary or if there are other changes that lower the CDF.

Another important consideration is whether and how well the licensee has addressed the human factors engineering aspects of the modification. The results of the human factors engineering analyses discussed in Sections 2, 3, and 4 must be considered in an integrated manner. No individual analysis is sufficient in and of itself. Thus, the decision will not be driven solely by the numerical results of the PRA. Each type of information helps in building an overall picture of the implications of the proposed change on risk. The PRA has an important role in putting the change into its proper context as it impacts the plant as a whole. As the discussions in the previous section indicate, both quantitative and qualitative arguments may be brought to bear. The different pieces of evidence that are used to make a final decision may not be combined in a formal way, but they do need to be clearly documented. The proposed change should be given increased NRC management attention when the calculated values of the changes in the risk metrics approach the criterion levels of current, accepted guidelines.

The main factors in the decision process are discussed here first and then supplementary decision factors are listed that may assist when the decision is difficult to make.

Main Decision Factors

- *Change in CDF* - One consideration is the value of ΔCDF_{mod} or the increase in Core Damage Frequency due to the modification. The placement of this value into the regions of Figure 2.1 can also be considered. The confidence one has in the PRA HEP value and hence that the change in CDF is at the value shown by ΔCDF_{mod} is partially determined by the results of the human factors review noted below.

- *Change in LERF* - Another consideration is Δ LERF, similar to CDF in #1 above.

- *Risk Importance Measures for the HA* - The values of RAW and FV give a measure of the risk importance of the HA in question. The specific meaning of these measures is discussed in Section 2. These provide insight on the contribution of the HA to risk.

- *Time and Integrated Risk* - A further consideration is the length of time that the change will be in place, if the modification temporary. The integrated risk over time (or the ICCDP and ICLERP) can be considered, per Section 2.

- *Human Factors* - A most important consideration is the degree of confidence that operators can perform the actions required for the modification in question. This is determined by the aggregate evaluation in Sections 3.2 through 3.12 of the Level I review guidance and Sections 4.2 through 4.4 of the Level II review guidance.

- *Deterministic Criteria* - Another consideration is the more traditional deterministic review guidance provided in Section 3.1 of the Level I review guidance and Section 4.1 of the Level II review guidance.

Supplemental Decision Factors

Additional factors may also be used, as appropriate, to determine the acceptability of a change. These were adapted from RG 1.174 Section 2.2.6, Integrated Decisionmaking and include:

- The cumulative effect of previous changes and the trend in CDF (the licensee's risk management approach)

- The cumulative effect of previous changes and the trend in LERF (the licensee's risk management approach)

- The effect of the proposed change on operational complexity, burden on the operating staff, and overall safety practices

- Plant-specific performance and other factors (for example, siting factors, inspection findings, performance indicators, and operational events), and Level 3 (offsite consequence) PRA information, if available

- The benefit of the change in relation to its CDF/LERF increase

- The practicality of accomplishing the change with a smaller CDF/LERF impact

- The practicality of reducing CDF/LERF when there is reason to believe that the baseline CDF/LERF are above the guideline values (i.e., 10-4 and 10-5 per reactor year)

6 REFERENCES

10 CFR 50. *Domestic Licensing of Production and Utilization Facilities.* Washington, D.C.: U. S. Government Printing Office.

ANSI/ANS. (1994). *Time Response Design Criteria for Safety-related Operator Actions* (ANSI/ANS-58.8-1994). La Grange Park, Illinois: American Nuclear Society.

Azarm, A., Higgins, J., and Chu, L. (1999). *Development of a Risk-informed Baseline Inspection Program* (BNL Report JCN W6234). Upton, New York: Brookhaven National Laboratory.

Higgins, J., O'Hara, J., Stubler, W., and Deem, R. (1999). *Summary of Credit of past Operator Action Cases* (Report No. W6022-T1-1-7/99). Upton, New York: Brookhaven National Laboratory.

Higgins, J., O'Hara, J., Lewis, P., Persensky, J. (2002). *Development of a Risk Screening Methodology for Credited Operator Actions,* In Proceedings for the 2002 IEEE 7th Conference on Human Factors and Power Plants. Phoenix, Arizona: IEEE.

Lambert, H. E. (1975). *Measures of Importance of Events* in *Reliability and Fault Tree Analysis,* ed. R. E. Barlow, J.B. Fussell, and N.D. Singpurwalla. Philadelphia, PA: SIAM Press.

NEI-00-02, *Probabilistic Risk Assessment (PRA) Peer Review Process Guidance.* Nuclear Energy Institute: Washington, D. C.

NRC (1980). *Criteria for Preparation and Evaluation of Radiological Emergency Response Plans and Preparedness in Support of Nuclear Power Plants* (NUREG-0654). Washington, D.C.: U.S. Nuclear Regulatory Commission.

NRC (1983). *Instrumentation for Light-water-cooled Nuclear Power Plants to Assess Plant and Environmental Conditions During and Following an Accident* (Regulatory Guide 1.97). Washington, D.C.: U.S. Nuclear Regulatory Commission.

NRC (1990). *Severe Accident Risks: an Assessment for Five U.S. Nuclear Power Plants* (NUREG-1150). Washington, D.C.: U.S. Nuclear Regulatory Commission.

NRC (1991). *Information to Licensees Regarding Two NRC Inspection Manual Sections on Resolution of Degraded and Nonconforming Conditions and on Operability* (Information Notice 91-18). Washington, D.C.: U.S. Nuclear Regulatory Commission.

NRC (1997). *Crediting of Operator Action in Place of Automatic Actions and Modification of Operator Actions, Including Response Times* (Information Notice 97-78). Washington, D.C.: U.S. Nuclear Regulatory Commission.

NRC (1997a). *Individual Plant Examination Program: Perspectives on Reactor Safety and Plant Performance* (NUREG-1560). Washington, D.C.: U.S. Nuclear Regulatory Commission.

NRC (1998). *An Approach for Plant-Specific, Risk-Informed Decisionmaking: Technical Specifications* (Regulatory Guide 1.177), August, 1998. Washington, D.C.: U.S. Nuclear Regulatory Commission.

NRC (1999). *SECY-99-007A, Recommendations for Reactor Oversight Process Improvements (Follow-up to SECY-99-007).* Washington, DC: U.S. Nuclear Regulatory Commission.

NRC (2000). *Technical Basis and Implementation Guidelines for a Technique for Human Event Analysis (ATHEANA)* (NUREG-1624, Rev.1). Washington, DC: U.S. Nuclear Regulatory Commission.

NRC (2001). *NRC Regulatory Issue Summary 2001-02: Guidance on Risk Informed Decision - making in License Amendment Revi*ews (RIS 2001-02). Washington, D. C.: U. S. Nuclear Regulatory Commission.

NRC (2002a). *NRC Regulatory Issue Summary 2002-22: Use of EPRI/NEI Joint Task Force Report, "Guideline on Licensing Digital Upgrades: EPRI TR-102348, Revision 1, NEI01-01: a Revision of EPRI TR-102348 to Reflect Changes to the 10 CFR 50.59 Rule."* (RIS 2002-22). Washington, D.C.: U.S. Nuclear Regulatory Commission.

NRC (2002b). *Human-system Interface Design Review Guidelines* (NUREG-0700, Rev. 2). Washington, DC: U.S. Nuclear Regulatory Commission.

NRC (2002c). *NRC Inspection Manual, Chapter,0609, Significance Determination Process, Appendix A, Attachment 2, Site Specific Risk-Informed Inspection Notebook Usage Rules.* Washington, DC: U.S. Nuclear Regulatory Commission.

NRC (2002d). *Standard Review Plan for the Review of Safety Analysis Reports for Nuclear Power Plants* (NUREG-0800), *Chapter 19, Use of Probabilistic Risk Assessment in Plant-Specific, Risk-Informed Decisionmaking: General Guidance.* Washington, D.C.: U.S. Nuclear Regulatory Commission.

NRC (2002e). *An Approach to Using Probabilistic Risk Assessment in Risk-informed Decisions on Plant-specific Changes to the Licensing Basis* (Regulatory Guide 1.174, Rev. 1). Washington, D.C.: U.S. Nuclear Regulatory Commission.

NRC (2003). *NRC Inspection Manual, Manual Chapter 0609, Significance Determination Process.* Washington, D.C.: U.S. Nuclear Regulatory Commission.

NRC (2004a). *Standard Review Plan for the Review of Safety Analysis Reports for Nuclear Power Plants* (NUREG-0800), *Chapter 18, Human Factors Engineering.* Washington, D.C.: U.S. Nuclear Regulatory Commission.

NRC (2004b). *Human Factors Engineering Program Review Model* (NUREG-0711, Rev. 2). Washington, D.C.: U.S. Nuclear Regulatory Commission

Thadani, A. (1998). NRC internal memo: Thadani to Collins, Report on Risk-Informed 50.59 Options, December 28, 1998. Washington, DC: U. S. Nuclear Regulatory Commission.

Vesely, W., Davis, T., Denning, R., and Saltos, N. (1983). *Measures of Risk Importance and Their Applications* (NUREG/CR-3385). Washington, DC: U.S. Nuclear Regulatory Commission.

GLOSSARY

Component - An individual piece of equipment such as a pump, valve, or vessel; usually part of a plant system.

Function - An action that is required to achieve a desired goal. Safety functions are those functions that serve to ensure higher-level objectives and are often defined in terms of a boundary or entity that is important to plant integrity and the prevention of the release of radioactive materials. A typical safety function is "reactivity control." A high-level objective, such as preventing the release of radioactive material to the environment, is one that designers strive to achieve through the design of the plant and that plant operators strive to achieve through proper operation of the plant. The function is often described without reference to specific plant systems and components or the level of human and machine intervention that is required to carry out this action. Functions are often accomplished through some combination of lower-level functions, such as "reactor trip." The process of manipulating lower-level functions to satisfy a higher-level function is defined here as a control function. During function allocation the control function is assigned to human and machine elements.

Human-system interface (HSI) - The means through which personnel interact with the plant, including the alarms, displays, controls, and job performance aids. Generically this includes maintenance, test, and inspection interfaces as well.

Human factors - A body of scientific facts about human characteristics. The term covers all biomedical, psychological, and psychosocial considerations; it includes, but is not limited to, principles and applications in the areas of human factors engineering, personnel selection, training, job performance aids, and human performance evaluation (see "Human factors engineering").

Human factors engineering (HFE) - The application of knowledge about human capabilities and limitations to plant, system, and equipment design. HFE provides reasonable assurance that the design of the plant, systems, equipment, human tasks, and the work environment are compatible with the sensory, perceptual, cognitive, and physical attributes of the personnel who operate, maintain, and support the plant (see "Human factors").

Mockup - A static representation of an HSI (see "Simulator").

Performance criteria - The criteria against which measured performance is compared in order to judge its acceptability. Approaches to the establishment of performance criteria include:

> *Requirement Referenced* - This is a comparison of the performance of the integrated system with respect to an accepted, quantified, performance requirement. For many variables a requirement-referenced approach can be used; i.e., requirements for plant, system, and operator performance can be defined through engineering analysis as part of the design process. Plant parameters governed by technical specifications and time requirements for important operator actions are examples of performance measures for which a requirement-referenced criteria can be determined. For performance measures where such specific requirement referenced criteria cannot be used alternative criteria development methods must be used.

> *Benchmark Referenced* - This is a comparison of the performance of the integrated system with that of a benchmark system which is predefined as acceptable under the same conditions or equivalent conditions. Such an approach is typically employed when no accepted independent

performance requirements can be established. Performance is evaluated through comparisons to an accepted benchmark rather that through an absolute measurement. For example, the evaluation may test whether the plant under review can be operated to stay within a level of operator workload not exceeding that associated with Plant X. Plant X is identified as acceptable for reasons such as its acceptable operating history and operators report their workload levels to be acceptable. In this case the performance measure must be obtained for Plant X and the new system, under similar operational conditions, and then compared. In the establishment of benchmark-referenced criteria, similar test conditions should be established for the benchmark system and system under evaluation.

Normative Referenced - Normative-referenced comparison is similar to a benchmark reference comparison, however, the performance criterion is not based upon a single comparison system, it is based upon norms established for the performance measure through its use in many system evaluations. The new system performs as compared to the norms established under the same conditions or equivalent conditions. This approach can be used when no accepted independent performance requirements can be established, but repeated use of the same performance measure enables the development of performance norms for acceptable and unacceptable systems.

Expert-Judgement Referenced - This is a comparison of the performance of the integrated system with criteria established through the judgement of SMEs.

Performance shaping factors (PSFs) - Factors that influence human reliability through their effects on performance. PSFs include factors such as environmental conditions, HSI design, procedures, training, and supervision.

Primary tasks - Those tasks performed by the operator to supervise the plant; i.e., monitoring, detection, situation assessment, response planning, and response implementation.

Risk-important human action - An action that must be performed successfully by operators to ensure plant safety. There are both absolute and relative criteria for defining these risk important actions. From an absolute standpoint, a risk-important action is one whose successful performance is needed to ensure that predefined risk criteria are met. From a relative standpoint, the risk- important actions constitute the most risk-significant human action identified.

Requirement - The term "requirements" is used in two different ways in this document: (1) requirements that are established as part of the design process; e.g., design requirements, functional requirements, task requirements, etc.; and (2) regulatory requirements identified in 10 CFR. There are no regulatory requirements established in this document.

Safety-related operator action - A manual action required by plant emergency procedures that is necessary to cause a safety-related system to perform its safety-related function during the course of any Design Basis Event. The successful performance of a safety-related operator action might require that discrete manipulations be performed in a specific order.

Secondary tasks - Those tasks that the operator must perform when interfacing with the plant, but are not directed to the primary task. Secondary tasks may include: navigating through and paging displays, searching for data, choosing between multiple ways of accomplishing the same task, and making decisions regarding how to configure the interface.

Simulator - A facility that physically represents the HSI configuration and that dynamically represents the operating characteristics and responses of the plant in real time (see "Mockup").

System - An integrated collection of plant components and control elements that operate alone or with other plant systems to perform a function.

Task - A group of activities that have a common purpose, often occurring in temporal proximity, and that utilize the same displays and controls

Testbed - The representation of the human-system interface and the process model used in testing.

Validation - The process by which the integrated system (consisting of hardware, software, and personnel elements) is evaluated to determine whether it acceptably supports safe operation of the plant.

Validity - The characteristics of the methods and tools used in the validation process. See the specific uses of the term: construct validity, convergent validity, performance representation validity, statistical conclusion validity, system representation validity, and test design validity.

Verification - The process by which the human-system interface design is evaluated to determine whether it acceptably reflects personnel task requirements and HFE design guidance.

Vigilance - The degree to which an operator is alert.

Workload - The physical and cognitive demands placed on plant personnel.

APPENDIX A

Generic Human Actions that are Risk-Important

This attachment contains two tables of generic HAs for BWRs and PWRs that are risk-important. Each table is further divided into "Group 1" HAs that are risk-important and "Group 2" HAs that are potentially risk-important. To facilitate readability of the tables, the names of common events and plant systems are given in acronyms. These acronyms are defined in the acronym list on page xiv of this document. These Tables are for use in the Generic Method described in Section 2.4 of this document.

Table A.1 Generic BWR human actions that are risk-important

Group 1: BWR Human Actions that are Risk-Important	
Human Actions	**Description and Reasons for Risk-Importance**
Perform Manual Depressurization	On selected sequences, such as station blackout (SBO), manual depressurization is required after failure of high pressure injection systems to allow for injection with low pressure systems. A complicating factor is that some procedures initially direct the operator to inhibit ADS. In some PRAs this appears in cutsets up to 45 % of CDF. Operators typically depressurize by manually operating the safety relief valves (SRV).
Vent Containment	On a transient or loss-of-coolant accident (LOCA) sequence, with failure of the PCS, containment temperature and pressure increase and must be controlled. This can be done by containment heat removal, suppression pool cooling, or containment venting. Actions are required to remove DH before adverse conditions are reached (e.g., high Suppression Pool temperature leading to loss of ECCS pumps).
Align Containment or Suppression Pool Cooling	
Actions During Shutdown	Almost all actions, including actuation of various equipment, are done manually during shutdown. The operator's understanding of the plant configuration is necessary for the successful manual actions.

Group 2: BWR Potentially Risk-Important Human Actions	
Human Actions	**Description and Reasons for Risk-Importance**
Level Control in anticipated transient without scram (ATWS)	Effective Rx Vessel level manual control at lower than normal levels (e.g., near the top of the active fuel) is needed during an ATWS in order to reduce core power.
Initiate Standby Liquid Control (SLC)	Manual initiation of SLC is needed for ATWS sequences.
Inhibit ADS	Some IPEs conclude that core damage will occur if ADS is not manually inhibited in an ATWS event due to instabilities created at low pressures.
Mis-calibrate Pressure Switches	Various pressure switches are important for initiating ECCS and operating ECCS permissives. Common cause mis-calibration of these switches can affect multiple trains of safety systems.
Initiate isolation condenser (IC)	For the early design BWR plants, this action is important during accidents to ensure the continued viability of the cooling from the IC.
Control feedwater (FW) events	The actions of operators to properly control the FW system as an injection source after loss-of-instrument air or other loss of FW events can be important in various sequences such as, transients and small LOCAs.

Recover Offsite Power	The actions of operators to recover offsite power after a total loss of offsite power (LOOP) is important to limit the risk due to station blackout and other LOOP core damage sequences. These are modeled with various recovery times in PRAs.
Shedding of DC Load After SBO	While often not well modeled, operator action to shed DC loads is needed to extend the battery charge in order to operate the AC independent HPCI and RCIC systems and to keep the SRVs open (to allow low pressure vessel injection from a diesel-driven fire pump). This extends the time to core damage and the time that operators have for recovery of AC power.
Similar actions to those in Group I	Actions that are substantially similar (but not identical) to those contained in Group 1 of this Table should be considered as potentially risk-important, if they involve the same systems, components, or actions.
Actions involving the risk-important systems	Each plant has a few systems that are clearly the most risk significant in the plant. Human actions associated with these systems should be considered as potentially risk-important. When modifications associated with these risk-important systems are being considered, new human actions may be created that were not in the original PRA, but that will be risk-important.

Table A.2 Generic PWR human actions that are risk-important

Group 1: PWR Human Actions that are Risk-Important	
Human Actions	**Description and Reasons for Risk-Importance**
Establish Recirculation	In LOCA scenarios, the switching of ECCS lines from the injection to the recirculation mode is done manually. Failure to do so or human error involving the valve alignment is important. Both low pressure and high pressure recirculation modes were noted to be important.
Feed and Bleed	Failure of the operator to initiate and perform the feed and bleed operation of the reactor coolant system as a last resort of heat removal is important. Of particular importance is the bleed portion using the pressurizer PORVs.
Provide Water Supply for AFW	Use of water pumps to transfer water, from other sources of make up to the CST for use by AFW, is considered important in scenarios when long term cooling through SG is needed.
RCP Trip	On a loss of cooling to the RCP seals, it is important for operators to quickly trip the pumps to prevent an RCP seal LOCA.
Action During Shutdown	Almost all actions, including actuation of various equipment, are done manually during shutdown. The operator's understanding of the plant configuration is necessary for the successful manual actions.
Group 2: PWR Potentially Risk-Important Human Actions	
Human Actions	**Description and Reasons for Risk-Importance**
Recover of RCP Seal Cooling	In some plants there are means of alternate cooling for RCP seals that could be relied on in scenarios involving loss of CCW. However, the alignment of the system is manual and requires operator action.
Recover Emergency AC or Offsite Power	Some losses of AC power can be recovered by either manual transfer of the source of power, or recovery of onsite normal/emergency AC power. This recovery action is considered risk significant in many PRAs.
Actions in Response to ATWS	Upon failure of RPS, the operator should perform several actions, starting with manual scram, ensuring turbine trip, and most importantly initiating emergency boron injection.
DEP and Equalization during SGTR Event	An important strategy during an SGTR event is the depressurization of primary and secondary systems and the equalization of pressures between primary and secondary. These all help to limit the leakage.
Isolate SG	During both a MSLB and an SGTR event, isolation of the affected SG is important.

Shut PORV Block Valve	During a stuck open PORV event, shutting the PORV block is an important action to eliminate the leak.
Isolate ISLOCA	In some plants there is a capability to isolate an interfacing systems LOCA through manual actions. Operator failure to isolate an interfacing LOCA in the LPI system is considered risk significant in these plants.
Similar Actions to those in Group I	Actions that are substantially similar to those contained in Group 1 of this Table should be considered as potentially risk-important, if they involve the same systems, components, or actions.
Actions Involving the Risk- Important Systems	Each plant has a few systems that are clearly the most risk significant in the plant. Human actions associated with these systems should be considered as potentially risk-important. When modifications associated with these risk-important systems, are being considered new human actions may be created that were not in the original PRA, but that will be risk-important.